WATER WORLDS:
HUMAN GEOGRAPHIES OF THE OCEAN

Water Worlds:
Human Geographies of the Ocean

Edited by

JON ANDERSON
Cardiff University, UK

and

KIMBERLEY PETERS
Aberystwyth University, UK

Routledge
Taylor & Francis Group

LONDON AND NEW YORK

First published 2014 by Ashgate Publishing

Published 2016 by Routledge
2 Park Square, Milton Park, Abingdon, Oxon OX14 4RN
711 Third Avenue, New York, NY 10017, USA

Routledge is an imprint of the Taylor & Francis Group, an informa business

British Library Cataloguing in Publication Data
A catalogue record for this book is available from the British Library

The Library of Congress has cataloged the printed edition as follows:
Anderson, Jon, 1973–
  Water worlds : human geographies of the ocean / by Jon Anderson and Kimberley Peters.
    pages cm.
  Includes bibliographical references and index.
  ISBN 978-1-4094-5051-1 (hardback: alk. paper)
  1. Human geography. 2. Oceanography. I. Title.
  GF41.A476 2014
  910.9162—dc23

                                                                    2013031531

ISBN 9781409450511 (hbk)

Contents

List of Figures and Tables

Figures

Table

Notes on Contributors

Jon Anderson is Senior Lecturer in Human Geography in the School of Planning and Geography, Cardiff University, UK. His research focuses on the relations between culture, place, and identity. He is particularly interested in the geographies, politics and practices that such relations produce. Jon has published widely in the fields of environmental action, qualitative methodology and most notably a textbook *Understanding Cultural Geography: Places and Traces* (Routledge 2010). Further information on his work can be found at www.spatialmanifesto.com.

Anyaa Anim-Addo is Lecturer in Caribbean History at the University of Leeds. She has research interests in the maritime world, the politics of mobility, and race and gender in the post-emancipation era. Publications include '"A wretched and slave-like mode of labour": slavery, emancipation and the Royal Mail Steam Packet Company's coaling stations,' *Historical Geography* 39 (2011), 65–84; and 'Steaming between the islands: nineteenth-century maritime networks and the Caribbean archipelago,' *Island Studies* 8:1 (forthcoming).

Christopher Bear is Lecturer in Human Geography in the School of Planning and Geography, Cardiff University, UK. His research focuses on human-animal-technology relations and geographies of knowledge and expertise. Much of his work has an aquatic focus and has encompassed studies of recreational angling, commercial fisheries and an aquarium. Most recently, he has completed an ESRC-funded project about the introduction of robotic milking technologies to the UK dairy sector.

Bärbel G. Bischof is a visiting Assistant Professor at Stetson University in Florida, U.S. Her research is focused on understanding the responses and connections of scientific discourse to constructions of environmental facts in ocean space. Framing ocean space in a geographic perspective in order to understand ecosystem management and human-nature interactions, she also serves as an Editorial Board Member of the *Journal of Ocean & Coastal Management.*

Juliette Hallaire is a PhD candidate at Keele University, UK, working on environment-induced migration, fishermen's mobility and maritime border experiences. She has been conducting in-depth field studies within local fishing communities in Senegal, developing a critical interdisciplinary approach between human geography, politics, mobility studies and maritime geographies. She graduated with a Masters in Geography at the University of Paris-IV/Sorbonne in 2007. From 2008 to 2010 she worked as a research assistant in migration-related programs in international organizations and NGOs in Colombia.

Anne-Flore Laloë works for the Marine Biological Association of the United Kingdom, developing and implementing projects in the field of ocean sciences, in conjunction with universities, schools and museums. Anne-Flore's research interests revolve around developing an interdisciplinary understanding of how knowledge about the Earth is produced and represented from scientific, cultural and social perspectives. Her doctoral thesis investigated the cultural and scientific practices that measured, represented and studied the Atlantic Ocean between 1492 and 1900. This work has been widely presented at international academic conferences including the 13th International Conference of Historical Geographers in Hamburg, and the Royal Geographical Society-Institute of British Geographers Conference in London.

Deidre Mackay is Senior Lecturer in Geography at Keele University, UK. The author of *Global Filipinos* (Indiana, 2012), she has longstanding research interests in migration, development and the geography of intimate ties, including those mediated by material culture and new communications technologies.

Stephanie Merchant is Lecturer in Sport (Social Sciences) in the Department of Education, University of Bath. Stephanie's current work explores the affective and therapeutic nature of seascapes with reference to skill development, health and wellbeing. Stephanie's research interests are primarily concerned with the body and technology, in particular sensual perception of land/seascapes, and multisensual video methods. Drawing on the works of Merleau-Ponty, Bergson, Deleuze, Haraway and Derrida her research has explored human(-technology)-nature relations in the contexts of Cultural Geography, Tourism and Education with recent publications in *Body & Society* and *Tourist Studies* and the book *Mediating the Tourist Experience* (Ashgate, 2013).

Kimberley Peters is Lecturer in Human Geography at Aberystwyth University, UK. Her research focuses on the intersections between place, mobility, material culture and the more-than-human, most recently in the context of the sea. Kimberley's work has been published in journals including *Area*, *Tourism Geographies* and *Environment and Planning A*.

Philip E. Steinberg is Professor of Political Geography at Durham University, UK. His research focuses on spaces whose geophysical and geographic characteristics make them resistant to state territorialization. His publications include *The Social Construction of the Ocean* (Cambridge, 2001), *Managing the Infosphere: Governance, Technology, and Cultural Practice in Motion* (Temple, 2008), *What Is a City? Rethinking the Urban after Hurricane Katrina* (Georgia, 2008), and *Contesting the Arctic: Politics and Imaginaries in the Circumpolar North* (I.B. Tauris, 2014).

Jonathan Taggart is a Vancouver-based photojournalist and PhD student at the Institute for Resources, Environment, and Sustainability at the University of British Columbia who specializes in social documentary, editorial photography, visual advocacy and visual ethnography (www.jonathantaggart.com). His work has been published in such journals as *Cultural Geographies*, *Environment and Planning A*, and *Transfers*.

Phillip Vannini is Canada Research Chair in Innovative Learning and Public Ethnography (www.publicethnography.net) and Professor in the School of Communication and Culture at Royal Roads University in Victoria, Canada. He is author/editor of nine books, including *Ferry Tales: Mobility, Place, and Time on Canada's West Coast* (Routledge, 2012) and *The Senses in Self, Society, and Culture* (Routledge, 2012).

Foreword
On Thalassography

'Seventy percent of our planet consists of oceans'. So began the proposal, drafted by Jon Anderson and Kimberley Peters, which ultimately resulted in this book. That figure is repeated continually in the ocean studies and marine environmentalism literature. Indeed, it is difficult to find a publication on endangered marine nature, the significance of global shipping, the role of fisheries in the food chain, the importance of the ocean for climate regulation, or the necessity of preserving maritime livelihoods, that does not contain this statistic or its more precise 71% variant.[1]

The statistic is a compelling figure. It achieves metaphysical significance when paired with the fact that the human body is also about 70% water. Indeed, it is so powerful that it may be having an unintended effect of, rather than instilling concern for our 'blue planet', leading to a sense of security: If so much of our planet is water, then why do we need to steward it as a fragile resource? Perhaps in response to this complacency, the United States Geological Survey (USGS) has recently complicated the statistical narrative with an image that demonstrates that while the planet's surface may be 70% ocean, the ocean, despite its depths, constitutes only .12% of the planet's volume (Figure F.1).

The USGS' reworking of the 70% figure illustrates how statistics often have multiple meanings. I would go one step further, however, and suggest that the prevalence of *both* the 70% statistic and the .12% image demonstrates how such figures can mask as much as they reveal. When one reduces the ocean – a dynamic system that is perpetually being remade and whose edges are continually being redefined – to a quantity (whether a seemingly large quantity like 70% or a seemingly small one like .12%), it becomes static and undifferentiated. The ocean can then be categorized as a space of nature to be fetishized, a space of alterity to be romanticized, or even a space beyond society to be forgotten. In each of these formulations, the ocean is classified as an *object*, a space of *difference* with a distinguishing ontological unity, the 'other' in a land-ocean binary.

---

1 In the interest of full disclosure, I confess that after writing this paragraph I looked back at my own early ocean works and there was the statistic, front-and-centre, in the first sentence of my article introducing the 1999 Geography of Ocean-Space focus section in *The Professional Geographer*: 'Despite its importance as a space crucial to the maintenance of many of the world's physical and social systems, the 71% of the earth's surface covered by the world-ocean has traditionally received much less attention from geographers than has the terrestrial sphere' (Steinberg 1999: 366).

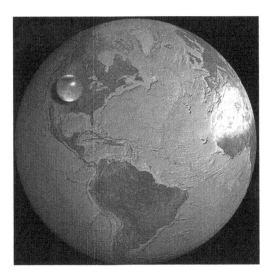

Fig. F.1   Volume of Earth's water compared to the total volume of Earth.

Source: United States Geological Survey (2012).

Note: The blue bubble depicts the total amount of water on earth – freshwater and biological water as well as seawater. However, since 96.54% of the planet's water is in the ocean, the volume of the bubble depicted is relatively close to the volume of a bubble that would represent only seawater. According to the USGS' figures, the volume of the ocean is 1,338,000 km^3, or .12% of the planet's volume, while the volume of all of Earth's water (which is what is represented by the blue bubble) is 1,386,000 km^3, or .13% of the planet's volume.

For the past two decades, critical theorists (including myself) have been problematizing the changing substance of that ontological unity, locating its shifting significations, uses, and regulatory regimes in the dialectics of capitalism, the interstices of political thought, or the specificity of cultural norms, and tracing its dynamism across time and space. But even as we have pursued this agenda, we have persisted in conceiving of the ocean as an object, a substance, a surface of difference, the *other* 70%. To borrow terms from Derrida (1982), in our efforts to identify *différence*, the system by which meanings are defined, we have ignored *différance*, the system by which meanings are deferred.

The alternative, if one is to write about the ocean as a non-objectified arena, is to approach it as a space that is not so much *known* as *experienced*; less a space that we live *on* (or, more often, gaze *at*) than one that we live *in*; less a two-dimensional surface than a four-dimensional sphere; a space that we think *from* (Anderson 2012).

However, there are two obstacles to applying this perspective to ocean-space. The first is that the ocean, as a material space, is particularly difficult to grasp. As Massey (2005) has demonstrated, a representation of space extracted from time obscures the social processes that constitute space (and time). Therefore, a point on land, as represented on a map, or, for that matter, in a planning document, is a false staticization of social processes. The same is certainly true for a point in the ocean. However, unlike a point on land (unless one thinks in geological time), the representation of a point in ocean-space is also a false staticization of *geophysical* processes. The ocean is constituted by vectors of movement – tides, currents, and waves – but these vectors do not simply occur *in* the ocean; they *are* the ocean (Steinberg 2011). As such, it is impossible to 'locate' a point at sea as an actual material place. Baudrillard's (2001) observation about the map preceding the territory is as true at sea as it is on land. But in the ocean there is a further iteration because the territory subsequently washes away the map. Thus we can never truly 'locate' ourselves within the ocean. Or, if we must locate ourselves, we require a different kind of 'map'.

And this connects to the second problem that emerges as we approach the ocean: Our encounters with the sea are always mediated. Whether by ships, scuba tanks, surfboards, or bodily movements, as well as, less physically, by stories, memories, sea shanties, fears, or dreams, our encounters with the sea are never 'pure'. There is always an outer layer between us and the sea that keeps us – and our experiences and thoughts – afloat. As countless philosophers, most notably Kant (1999), have shown, this is a problem endemic to humans regardless of their environment. But it is particularly profound at sea, where both survival and interpretation require reliance on resources that we are aware we have borrowed from somewhere else.

Since the sea is a space that cannot be located and that cannot be purely experienced, thalassography – sea-writing – presents a challenge.[2] It is no wonder that the social science literature on the sea as a holistic space of interspecies intersubjectivity is exceptionally sparse; in many ways the ocean seems to be a space more suitable for the literary essay or poem that reproduces difference even as it interrogates its foundations, for the policy analysis or military strategy that analyzes one particular ocean use while ignoring others, or for the philosophical tome that reduces the sea to a metaphor for flux and flow while ignoring the actual mobilities that are experienced by those who traverse or gaze upon its surface. In our efforts to understand the ocean as an arena characterized by the co-construction of maritime subjects – from sailors and swimmers to reefs and water molecules – we geographers face a formidable task.

---

2   In the nineteenth century, the term 'thalassography' referred to the science of marine organisms (see, for instance, Agassiz' (1888) *Contribution to American Thalassography*). The term has fallen into disuse among oceanographers and marine biologists, however, so now seems an opportune time to reclaim it as an aquatic analog to geography (earth-writing). A similar term – 'thalassology' – has been adopted by some in the Mediterranean studies community for the study of ocean-centred regions (Horden & Purcell 2000, 2006; Mentz 2009), but, as I discuss elsewhere, the study of ocean-centred regions is not strictly the same as the study of ocean encounters (Steinberg 2012).

The brilliance of the contributions in this book is that each author perceives the obstacles inherent in narrating stories of the sea, but each also recognizes that advances in social scientific thought, in the discipline of geography and beyond, provide new tools for interpreting how the ocean is experienced through a prism of historic and ongoing material encounters and discursive understandings. Indeed, the contributors to this collection view the barriers to effective thalassography as opportunities for applying and expanding some of the most innovative tools in social thought.

Fifteen years ago, when I authored the introduction to *The Professional Geographer*'s Geography of Ocean-Space focus section, I followed my reincantation of the 70% statistic with the statement, 'If indeed there ever was a valid excuse for lack of geographic research in marine space, that time has certainly passed' (Steinberg 1999: 367). I justified this statement with references to the sea's crucial role in the world economy, regional livelihoods, and cultural imaginaries. However, although I failed to realize this at the time, human geographers still had a valid excuse for ignoring the sea: We lacked the conceptual and analytic tools for grasping this exceptionally ungraspable space. That situation has changed considerably since 1999, and I am proud to join the *Water Worlds* editors and contributors in launching a new wave of thalassography.

Philip E. Steinberg

References

Agassiz, A. 1888. *A Contribution to American Thalassography: Three Cruises of the United States Coast and Geodetic Survey Steamer "Blake" in the Gulf of Mexico, in the Caribbean Sea, and along the Atlantic Coast of the United States from 1877 to 1880*, 2 vols. Boston: Houghton, Mifflin.

Anderson, J. 2012. Relational places: The surfed wage as assemblage and convergence. *Environment and Planning D: Society & Space*, 30, 570–87.

Baudrillard, J. 2001. Simulacra and simulations. In *Jean Baudrillard, Selected Writings*, 2nd ed., edited by M. Poster. Stanford, CA: Stanford University Press, 169–87.

Derrida, J. 1982. Différance. In *Margins of Philosophy*, by J. Derrida, trans. A. Bass. Chicago: University of Chicago Press, 1–28.

Horden, P. & Purcell, N. 2000. *The Corrupting Sea: A Study of Mediterranean History*. Oxford: Wiley-Blackwell.

Horden, P. & Purcell, N. 2006. The Mediterranean and "the new thalassology". *American Historical Review*, 111, 722–40.

Kant, I. 1999. *Critique of Pure Reason*, trans. P. Guyer and A.W. Wood. Cambridge, UK: Cambridge University Press.

Massey, D. 2005. *For Space*. London: Sage.

Mentz, S. 2009. Reading the new thalassology. In *At the Bottom of Shakespeare's Ocean*, by Steve Mentz, London: Continuum, 101–12.

Steinberg, P.E. 1999. Navigating to multiple horizons: Toward a geography of ocean-space. *The Professional Geographer*, 51(3), 366–75.

Steinberg, P.E. 2011. Free sea. In *Sovereignty, Spatiality, and Carl Schmitt: Geographies of the Nomos*, edited by S. Legg. London: Routledge, 268–75.

Steinberg, P.E. 2013. Of other seas: Metaphors and materialities in maritime regions. *Atlantic Studies*, 10(2), 156–69.

United States Geological Survey. 2012. Water science photo gallery: how much water is on Earth [Online, updated 18 May] Available at: http://ga.water.usgs.gov/edu/2010/gallery/global-water-volume.html.

'He had bought a large map representing the sea,
Without the least vestige of land:
And the crew were much pleased when they found it to be
A map they could all understand.
...
"Other maps are such shapes, with their islands and capes!
But we've got our brave Captain to thank:
(So the crew would protest) "that he's bought us the best –
A perfect and absolute blank!"'
(The Hunting of the Snark, Lewis Carroll, 1876).

INTRODUCTION

Chapter 1

'A perfect and absolute blank'
Human Geographies of Water Wc

Jon Anderson and Kimberley Peters

Introduction

Our world is a water world. The oceans and seas are entwined, often invisibly but nonetheless importantly, with our everyday lives. Trade, tourism, migration, terrorism, and resource exploitation all happen in, at, and across the oceans. The globalized world of the twenty-first century is thus thoroughly dependent upon water worlds. Despite this, geography, as 'earth writing' (Barnes and Duncan 1992: 1), has largely taken its etymlogical roots seriously (Steinberg 1999a, Peters 2010). The discipline has been a de facto terrestrial study; the sea not accorded the status of a 'place' worthy of scholarly study (Hill and Abbott 2009: 276). In the words of Lewis Carroll's crew in *The Hunting of the Snark* (see Foreword), until very recently, geography has reduced the sea to 'a perfect and absolute blank'. Such status has been most marked within *human* geography, where focus on socio-cultural and political life rarely strays beyond the shore (Steinberg 1999a: 367). As Mack identifies, water worlds have generally been relegated to,

> either ... the backdrop to the stage on which the real action is seen to take place – that is, the land – or they are portrayed simply as the means of connection between activities taking place at coasts and in their interiors. (2011: 19)

As a consequence, the predominant view of the sea has come to be characterized as,

> a quintessential wilderness, a void without community other than that temporarily established on boats crewed by those with the shared experience of being tossed about on its surface. (Mack 2011: 17)

Such a conceptualization is commonly attributed to 'modern' framings in the industrial capitalist era that have endured until the twenty-first century (Steinberg 2001: 113). Oceans and seas have been dismissed as spatial fillers to be traversed for the capital gain of those on land (Steinberg 2001) or conquered for means of long distance imperial control (Law 1986, Ogborn 2002). Moreover, because so few moderns live their lives at sea – it is not a place of 'permanent, sedentary habitation' (Steinberg 1999a: 369) – water worlds often remain at the edge of everyday consciousness. As Langewiesche states,

> Since we live on land, and are usually beyond the sight of the sea, it is easy to forget that our world is an ocean world. (2004: 3)

Accordingly, within human geography, greater interest has been paid to the land: to cities, towns, streets, homes, work places, leisure centres, schools – the places which are *seen* to be crucial to our everyday existence (Peters 2010: 1263). Furthermore, according to Steinberg, the marginalization of the maritime world is further compounded due to difficulties researchers face in accessing areas of the sea which are inhospitable, detached from the shore, physically unstable and immensely deep (1999a: 372). This inaccessibility has resulted in a vision of water worlds, projected by scholars, artists and writers, which is abstracted and distanced from reality. As Steinberg puts it, 'the partial nature of our encounters with the ocean necessarily creates gaps' in how the ocean is understood (2013: 157). Consequently, the physical liveliness of oceans and seas are often reduced to romantic metaphors in paintings, novels and other literary and art sources. Together, these reasons have resulted in a largely 'landlocked' discipline (Lambert et al. 2006: 480).

However, over the past decade, geographical research has cast off its terrestrial focus and has begun to voyage towards new, watery horizons. This book brings together scholars concerned with the manifold human geographies of the sea, acting as a first 'port of call' for those interested in taking research offshore, as well as offering exciting new theoretical and empirical in(ter)ventions in thinking about our water world. This book contends, along with Lambert el al (2006), that water worlds must move from the margins of geographical consciousness and inquiry (see also Peters 2010, Steinberg 1999a, 1999b, 2001). This means, to echo Steinberg in the Foreword to this volume, that we must not simply study the seas and oceans as 'other' or 'different' spaces; but instead start thinking *from* the water. With this in mind, this book aims to chart new representations, understandings and experiences of the sea, plotting water worlds that are more than a 'perfect and absolute blank'.

To this end, the book has three main aims. Firstly, to shift the sea to the centre of human geographical studies. No longer, we contend, can the sea be conceptualized as marginal to the land (and thus less significant to our everyday existence) and nor can it be positioned as peripheral to our academic enquiries, inferior to the terrestrial studies of 'landed' socio-cultural and political phenomena. Secondly, and relatedly, we seek to demonstrate the ways in which the sea is not a material or metaphorical void, but alive with embodied human experiences, more-than-human agencies and as well as being a space in and of itself that has a material character, shape and form. Finally, we propose that attention to oceans and seas may open up a new way of thinking, not only about these particular spaces, but also beyond them, to our terrestrial and aerial worlds too. Here we suggest a shift towards a 'fluid ontology', promoting a knowledge of the world which is *neither* 'land' biased *nor* 'locked' to static and bounded interpretations of space, but rather one that conceives of our (water)world as one which is in flux, changeable, processual and in a constant state of becoming. In the remainder of this chapter, we steer a course through each of these aims, attending to them in greater detail, before outlining the shape of the book and chapters which follow.

(Re)centering Water Worlds

The enduring marginalization of the seas and oceans from much human geographical inquiry has led to key lacunae in our understanding of the contemporary and historical world. It is the longstanding binary between the 'land' and 'sea', whereby the latter category is afforded an inferior status to the former, which has cemented the position of water worlds as 'outside' of academic study (see Shields 1992). As Westerdahl points out, there is a commonly held binary which separates the land and sea (2005: 13), negatively coding the ocean as 'different' or 'other' (Jackson 1995: 87–8). As Jackson notes,

> … geographers … bound by a European terrestrial bias, have accepted as *natural* the dominance of the land in understanding human interactions and relationships with environments. (1995: 87–8 original emphasis)

This 'naturalized' position of the oceans as marginal to the land, is, moreover, enforced through the liquid materiality of water. The sea's physical constitution renders it as intrinsically 'other'; it is a fluid world rather than a solid one. Our normative experiences of the world centre on engagements *on* solid ground; rather than *in* liquid sea (although see Craciun 2010, and Vannini and Taggart, this volume, who complicate this binary). As such, the watery composition of (most) seas and oceans mean they cannot be populated, with material manifestations and human life, in the same way as the land. Consequently, through such material difference, such spaces are visually different also. The material shape and form of the sea is 'other' to the stability, and therefore to the aesthetic features (both natural and humanly constructed), which characterize the land. The sea precludes the easy development of buildings and structures and covers the intricate landscape which lies under its surface in the shape of the seabed. As such, the difference between landed and watery realms is often difficult to interpret and understand from a strictly modern perspective. As anthropologist Levi-Strauss reflected, during a voyage at sea,

> I feel baulked by all this water which has stolen half my universe … The diversity customary on land seems to me to be simply destroyed by the sea, which offers vast spaces and additional shades of colouring for our contemplation, but at the cost of an oppressive monotony and a flatness in which no hidden valley holds in store surprises to nourish my imagination … the sea offers me a diluted landscape. (1973: 338–9)

For Levi-Strauss then, the land is 'full' of a variety of natural features that create a rich topography for the spectator. The sea, in opposition, is empty of such variety: it is flat, 'monotonous', and 'oppressive'. Subsequently, from his landed gaze, the sea is relegated to an inferior position within a binary to terra firma. Such a view of the ocean is not uncommon and has underscored the omission of oceans and seas from geographical study. It is only in the discipline's more recent history that the study of water worlds have begun to surface (see Steinberg 1999a, 2001, Lambert et al. 2006, Peters 2010).

Yet whilst there is evidence of engagement with watery worlds in much emerging literature in human geography (and related disciplines), the sea continues to be positioned as a subsidiary concern, rather than a central one. According to Peters (2010), the inferior position of the sea is compounded when scholars do not consider it as a material space with its own narrative, but rather employ it as a means to explore other socio-cultural phenomena. An example of this is Paul Gilroy's use of the 'Black Atlantic'. Gilroy uses the image of a 'Black Atlantic' ocean to reconceptualize and understand the relationships between nation, race and ethnicity which have been typically treated as cultural absolutisms (1993: 3). Thus, in Peters' words, rather than enhancing our understanding of the sea, such work operates in,

> the re-visioning of objects, themes and sites of study within a frame of the 'maritime' and 'oceanic' [only serving to] aid a greater understanding of the workings of, for example; ... colonial expansion, empire, and ... "historical master narratives" of the nation state. (Lambert et al. 2006: 480) (Peters 2010: 1261–2)

Although using the sea as a conceptual device to understand such processes is an important objective, such studies nevertheless serve to reinforce the apparent superiority of landed life to the detriment of investigation into the sea in and of itself. To be clear, this is not to advocate that water worlds are taken as a 'perfect and absolute' bounded space to examine in opposition to the attention paid to the land. Indeed, much of the richness of recent work that has incorporated the sea demonstrates how water worlds are spaces across which new connections, knowledges and experiences are realized (see Armitage and Braddick 2002, Featherstone 2005, Lambert 2005, Ogborn 2002). Indeed, the sea is a space intrinsically connected to and absorbed within a broader network of spaces (earth and air) which are also, likewise, porous, open and convergent with each other. However, we do argue that oceans and seas are recognized as equally fundamental *within* processes of socio-cultural, political and economic transformation, rather than acting merely as conceptual devices for understanding those processes. Accordingly, we contend (along with others, Steinberg 2001, Lambert et al. 2006) that where the seas feature in scholarship, they are not merely present as a secondary concern, but are fully folded into geographical research, 'demonstrating the potential – perhaps even freedom – offered by the sea' (Lambert et al. 2006: 480).

Indeed, since 2001 (with Steinberg's book *The Social Construction of the Ocean*), this effort is now fully underway. Social and cultural geographers have increasingly recognized the absence of maritime worlds in scholarly discussion and have worked to fill this liquid void, demonstrating it as far from the monotonous, empty plain of Levi-Strauss' landed view, or a mere conceptual tool for understanding (often) non-watery phenomena. Rather, viewing the sea *from the sea* offers a far more nuanced and complex perspective on the sea itself and the merits of developing a geography *of the sea* in which the sea takes centre stage. As Raban writes, the sea holds much character and complexity for contemplation in its own right:

> Seen from the cliffs, the sea might have looked as evenly arranged as the strings on a harp – the lines of white-caps running parallel at intervals of sixty feet or so. Seen from the wheel of a small boat, it presented quite a different aspect. Each wave in the train carried a multitude of smaller deformities – nascent waves bulging, heaping, trying to break as they rode the back of the senior wave in the system. (1999: 165)

This emerging interest in human geographies of the ocean has considered, in manifold ways, the spatialities bound up in and through water worlds. Studies have developed through a variety of lenses, including, for example, the networks of flows across ocean spaces, the study of specific maritime communities (sailors for example), the exploration of maritime places, such as the port, or the ship and the ways in which some non-modern cultures have the water as central to their world (see Peters 2010 for a review). Thus, as Lambert et al. identify, in recent years increasing attention has been paid to 'epistemological and historiographic perspectives ... the imaginative, aesthetic and sensuous ... and ... material and social geographies' of the oceans' (2006: 480).

Such a move is justified when we consider the influence of the sea on our everyday lives. As Lavery tells us, in contemporary society, approximately '95% of trade is still carried by ship' (2005: 359). Gifts for Western celebrations arrive freighted by sea from Asia; the global need for oil is serviced by giant tankers exporting resources from the Middle East to far flung ports; whilst modern day piracy on the high seas raises the costs of goods and insurance premiums, felt in consumers' pockets across the globe. Such phenomena alert us to the mobilities across the water that permeate and infiltrate our daily existence in often unnoticed, but highly significant ways. No longer then, should we think of water worlds as empty of activities, mobilities and lifeworlds. The seas are tied up with, and intrinsic to, a host of social, cultural, economic, political and environmental questions.

In this book, our studies launch from the starting point that seas *are* significant. In Cooney's words, we envision a study of the sea that is,

> contoured, alive, rich in ecological diversity and in cosmological and religious significance and ambiguity – [providing] a new perspective on how people... actively create their identities, sense of place and histories. (Cooney 2003: 323)

Through each of the chapters that follow, the authors in this collection assert that we should consider the sea not as a space defined in negative relationality to the land, but as central to processes of knowledge production, embodied experience and to understanding the more-than-humanness of our world. Firstly, therefore, this book aims to continue to establish a human geography of the ocean which takes the water itself, and its *central* connections to the land, seriously.

Filling the Watery Void

The second key aim of this volume is to continue challenging the aforementioned and long standing configuration of the ocean as an empty space, established through processes of industrial and post-industrial capitalism. Moreover this book seeks to address the use of the sea as a mere conceptual device for understanding alternative socio-cultural and political phenomena, instead positioning the sea as a 'an element of nature itself' (Steinberg 2001: 167). As such, the chapters which follow each demonstrate the ways in which ocean is 'filled': through its own elemental composition, with more-than-human life, with floating and sunken materialities, and with a range of human significance.

Such an approach takes inspiration from non-Western perspectives of the water world. If we turn away from our modern, terrestro-centric view, we can begin to see how 'other' cultures conceive of the seas and oceans as practiced, embodied and lived spaces. For example, anthroplogist Bronislaw Malinowski demonstrates the importance of rituals at sea for societies on the Trobriand Islands in the Western Pacific region (1922). Here the land functions as a connection point whilst the ocean is encultured as a significant ritualized space, made meaningful through the 'Kula' system of gift-giving. Kula exchanges involve the sea-based exchange of two types of item (armshells and necklaces) between 'Kula partners' (Young 1979: 163). Articles are moved from island to island by sea-going canoes, and as such, seafaring has been integral to the custom, culture, and ceremony of the Trobriand people (Young 1979: 172–3). Thus despite Western culture's willingness to reduce the water world to an empty space, many 'indigenous' cultures refute this essentialism. As Raban notes, drawing on David Lewis' discussion of Polynesian mariners, *We, the Navigators* (1994):

> the open sea could be as intimately known and as friendly to human habitation as a familiar stretch of land to those seamen who lived on its surface, as gulls do, wave by wave. … the stars supplied a grand chart of paths across the known ocean, but there was often little need of these since the water itself was as legible as acreage farmed for generations. Colour, wind, the flight of birds, and telltale variations of swell gave the sea direction, shape, character. (Raban 1999: 94)

According to Raban, this intimate knowing of water worlds was supplanted in the West by the advent of modern technology, starting with the use of a compass and sextant and extending through to twenty-first century exploitation of satellite telemetry and geographical positioning systems. For Raban, 'the arrival of the magnetic compass caused a fundamental rift in the relationship between man [sic] and sea' (1999: 95). Possession of a compass,

> rendered obsolete a great body of inherited, instinctual knowledge, and rendered the sea itself – in fair weather, at least – as a void, an empty space to be traversed by a numbered rhumb line. (1999: 97)

Yet, as this volume demonstrates, there remains an embodied knowledge waiting to surface in Western (as well as non-Western) contemporary engagements with the water. Many individuals and cultures now understand and experience the sea as a 'place' with character, agency and personality (see Laloe, Anderson, Merchant, Hallaire and McKay, this volume). As Anderson explains with respect to surfing practice, when encountering and riding a wave, boarders experience 'stoke', a '"feeling of intense elation"', '"a fully embodied feeling of satisfaction, joy and pride"' (2012: 576, citing Evers 2006: 229–300). As such, this volume examines how humans do not just imagine water worlds, they actively engage with them in a wholly embodied way. Such embodied practice with water makes possible the writing of new corporeal experiences, impossible to fathom through landed, grounded explorations alone.

In the broader social sciences there is a growing recognition that embodied experiences of the world are integral to both our humanity and understanding (see Davidson, Bondi and Smith 2005). Emotions and more-than-cognitive understandings (see Pile 2010) are therefore seen as increasingly essential components in our knowledge systems, as well as inevitable productions from our interactions with the (water) world of which we are a part. It is through affects and emotions that we 'literally make sense of the world' (Wood and Smith 2004: 534). Accordingly, in this book we draw on theories which enable us to engage with the practices and performances through which we encounter the world. To this end we recognize that representation can only take us so far in knowing water worlds. As Thrift tells us, 'the varieties of stability we call 'representation' can only cover so much' (2004: 89), thus it is vital to consider how the seas and oceans are thoroughly more-than-representational (after Lorimer 2005) in nature.

In thinking of water worlds as more-than-representational spaces, we can be alerted to the many ways in which seas and oceans 'come to life'; the non-human actors, materialities and natural states of water which all merge in this processual and fluid medium. Indeed, in this book we not only seek to draw attention to the activities and embodied practices made possible at sea to reveal new visceral knowledges, we also contemplate the role of non-human actors that fill this void: the fish, insects and rodents (see Bear and Eden 2010, and Bear and Anim-Addo, this volume) and multiple materialities which reside, on, in, and under the oceans: ships, surf boards and even trucks (see Anderson, Merchant and Vannini and Taggart, this volume).

Such a move also echoes broader steps in the social sciences to think beyond a world simply constructed by humans. As Bennett writes, 'humans are always in composition with nonhumanity' (2004: 365). Thus human geographies of water worlds require us to think seriously about the water itself as a non-human materiality (Jones 2011, Peters 2012). In a manifesto for a return to the 'livingness' of the world, Whatmore contends there is a need to 're-animate the missing matter of landscape' (2006: 605). Yet arguably, we must also recognzse the 'missing matter of *seascape*' (Peters 2012: 1242), and thus in this book we must pay attention to the very nature of the sea itself.

Indeed, the physical quality of the sea (in liquid form) makes it a mobile medium subject to the energies and forces of nature – the wind, jet streams, the extra-terrestrial gravitational pull of the moon. As Jones notes, 'ocean rhythm patterns are...expressions of the interplay of many profound forces' (2011: 2287), which, in turn, make it a volatile, undulating, dynamic three dimensional materiality (Peters 2012: 1242). Furthermore, in solid form, or as gaseous matter, the water world takes on differing states which open up alternative possibilities for co-constitution with human life and experience (see Vannini and Taggart, this volume). As such, this book moves beyond the oceans as a flat, empty space, or one of only abstract representation, to a space that is living, and has its own agencies. It urges us to consider the various *lives* that 'fill' ocean space and the manifold *things* that surface 'on' or 'within' the sea, as well as taking seriously the very *matter* of water itself.

Considering the sea as a space which can be 'known' (Part I), a space which is viscerally experienced (Part II) and which has its own nature (Part III) the chapters of this book demonstrate a new way of conceptualizing the seas and oceans which move them from the margins of human geographical concern. In starting from the sea (rather than the land) and drawing from both cognitive and actual engagements with oceans, this book takes it final step in seeking to develop a new *fluid* language to understand our water world; a language that is necessary if we are to take the human geographies of the oceans seriously.

Towards a Fluid Ontology

To paraphrase Cresswell (2000: 263), we have argued to date that human geographers have yet to adequately 'talk about the water'. Despite this, in various ways the marine and maritime world lap into our everyday lives through the use of language. Sayings such as 'all hands on deck', 'you can't swing a cat in here' or 'all at sea', have moved seamlessly from ship-based contexts to land-based life, and thus, turns of phrase are appropriated and tie together the terrestrial and water world. Such sayings remind us that language remains a key way through which humans make sense of our geographies. As Wittgenstein reminds us, 'the world we live in is the words we use' (cited in Raban 1999: 151), and although 'how places are made is at the core of human geography, [we have perhaps] neglected the explicit recognition of the crucial role of language' in creating it (after Tuan 1991: 684). Sensitizing ourselves to language, and the philosophies that underpin it, is therefore a further means by which we can take the 'large [blank] map representing the sea' and reinvent it as something we can 'all understand' (see Carroll's *The Hunting of the Snark*, Foreword).

As our world is not only terrestrial but also marine in focus, it is possible to move beyond a traditional terrestrial vocabulary that is dominantly used to describe, conceptualize, and understand it. According to Cresswell (2006), the dominant way in which the terrestrial world is studied is through adopting the language of a 'sedentary metaphysics' (after Malkki 1992). This language seeks to 'divide the world up into clearly bounded territorial units' (Cresswell 2004: 109), whilst the process of place-making involves the 'carving out of *permanences*'' (Harvey 1996: 294, emphasis added). It is from this sedentary metaphysics that our 'common sense' categorizations of the world – as fixed, static and durable – originate (Bourdieu 1977, 1991). Although this narrative may be appropriate for many places, it also produces a limited geographical imagination in a number of ways. Due to the radical difference in physicality between the terrestrial and the oceanic (Steinberg 1999a: 327), these perspectives serve to not only marginalize the marine world from scholarly study, but also preclude theoretical innovations that may help to conceptualize this world more appropriately. We have seen how geography has always been a 'land' discipline, but in this way is also became a 'locked' discipline, fixated on the sedentary, static and terrestrially rooted rather than processes of flow, hybridity and mobile routes.

Jettisoning a sedentary metaphysics questions the imposition of clear, stable ontological categories onto the world. In a world of flow, change, and hybridity, products are rather seen as processes that have only temporarily stabilized. Movement and mobility is primary, there is a recognition that 'things' are simply pauses in the process of becoming something else. In the words of Dovey, this world of immanence rejects,

> the Heideggerian ontology of being-in-the-world [and replaces it] with a more Deleuzian notion of becoming-in-the-world. This implies a break with static, fixed, closed and dangerously essentialist notions of place, but preserves a provisional ontology of place-as-becoming: there is always, already and only becoming-in-the-world (2010: 6).

At a superficial level, such a shift in metaphysics suits the sea. The sea is obviously fluid; it is moving in terms of its location, it is unstable in terms of its form (from still calm to waves, to tides, to storm surges and tsunamis), and changeable in terms of its chemical state (as either solid (ice), liquid or water vapor). The water world is therefore in a constant state of becoming, it is a world of immanence and transience. The water world has a fluid ontology.

If, as Melville describes, we live in a '"terraqueous globe"' (cited in Philbrick 2001: 235) which is formed through an assembled mix of land and ocean (and air), it is appropriate to question how these fluid ontologies effect our understandings of this assemblage (see DeLanda 2006, Deleuze and Guattari 2004, Dovey 2010) What is the most appropriate language to adopt in our conceptualization of this world, and what consequence might it have for our understanding of it? As Vannini (this volume) asks: 'What would happen if we instead viewed the land from the perspective of water? What would happen if we wanted to see similarities and overlaps between land and water, rather than distinctions and boundaries?' A move towards a fluid ontology of the oceans is thus not to claim that water worlds are taken as a perfect and absolute *bounded* space to study, in opposition to the attention paid to the land. Rather, a fluid metaphysics alerts us also to the ways in which the land and air fluidly merge and mix with water worlds too. In other words, it allows us to consider the equally fundamental role of water which is intrinsically *connected to* and *absorbed within* a broader network of spaces (earth and air) which are also, likewise, porous, open and convergent with each other. Understanding the oceans and seas as spaces intertwined with land, air and human life allows us to move beyond conceiving them as empty, but rather as part of a 'meshwork' of natures integrated into human experience (Ingold 2008).

This fluid ontology thus offers us a new perspective from which to rethink the constitution of the world. As Anderson, drawing on Goldsworthy, notes in Chapter 7, maybe the world is fluid (see also Strang, in Vannini, this volume). Perhaps by this 'simple' change in vocabulary we can rethink our 'earth writing'. We can start from the assumption that the world is becoming and look at the ways things become relatively stable, rather than the other way around. We can start from a state of transience and fluidity, and trace how states become more permanent and durable. Such a task thus contributes to Deleuze's call to overturn the 'privileging of stable [states as] a central tenet of Western metaphysics' (in Dovey 2010: 22). In its place, we begin to consider how fluid ontologies are temporarily stabilized in their mobility, trajectory and constitutional state (e.g. solid, liquid or gas).

Following this argument, we can accept that the sea (and even the world in general) is not static or stable, it is only of the immediate present, before it becomes something else. It is on a line of mobility and flow that it constantly taking it elsewhere. Our fluid ontology is forever emerging and emergent. Although the water world is more than 'just a metaphor' we may nevertheless turn to its representational function to facilitate a shift in language and philosophy to better understand our place in it. As Tally puts it:

> The human condition is one of being 'at sea' – both launched into the world and somewhat lost in it – and, like the navigator, we employ maps, logs, our own observations and imagination to make sense of our place. … The experience of being in the world is one of constant navigation, of locating oneself in relation to others, of orientation in space and in time, of charting a course, of placement and displacement, and of movements though an array of geographical and historical phenomena (2013).

Book Outline

This book brings together a collection of chapters by key authors, whose research interests focus explicitly on theorizing *from* the oceans and seas. Through these chapters a fluid approach to studying the sea is advocated, not only theoretically (as outlined above) but practically, in view of how we approach sea-based studies. Indeed, in what follows, the chapters move fluidly from human geography to cognate disciplines including maritime history, cultural studies and environmental science. Accordingly, this volume also incorporates a series of empirical interventions which endorse a breadth of methodological approaches – quantitative surveys, archive material, textual analysis of magazines, films, documentaries and books to interview and focus group data. The chapters move fluidly across epochs, from the 1500s to the present day, and focus our attention on both elite and everyday used of water worlds. Moreover, this book moves scholarship beyond an Atlantic-centric bias, presenting the global connectedness of the oceans as the chapters move across and between the Caribbean and North Seas; the Indian, Pacific, Atlantic and Arctic Oceans; the Gulf of Mexico and the Mediterranean. It also shifts the bias of thinking of oceans and seas as watery – considering how the water world fluidly infiltrates between different material 'states' which can be liquid, but also solid (as liquid solidifies to become ice) and air (as liquid evaporates).

This volume also takes us beyond the surface of the water to consider oceans and seas as three dimensional material spaces. The materiality of water worlds (in liquid form at least) means that we may be immersed in and 'converge with' them in distinct ways. It also encourages us to follow or trace the fluid routes of water beyond the ocean, towards connected bodies of water, rivers, lakes and streams. Indeed, this collection further explores the fluid mobilities of water worlds *beyond borders,* encouraging us to think of seas and oceans as intrinsically part of a larger assemblage forged in connection with the land, air and extra-terrestrial forces (such as the moon). It inspires us to think also beyond the borders of anthropocentrism – of human relations with water and water as socially constructed – to instead think about those more-than-human actors and the affects which arise when water and life coalesce. This leads us to begin to fluidly move beyond a landed language, a sedentary metaphysics, to instead reconfigure how we conceptualize our world more broadly.

The three parts of the book tack a route through various ways in which geographical work can start *from* the seas and oceans, producing 'water writing', rather than just 'earth writing'. Part I considers knowledge which can be gained *from* the seas and oceans, exploring how such knowledge unlocks new ways of thinking about our world, not possible from a terrestrial focus. These chapters offer fresh perspectives on how we make sense of political, historical, environmental and cultural concerns which shape our understandings of the world more broadly. Additionally, each of these contributions offers a differing take on the kinds of knowledge an ocean-centred approach reveals; from scientific knowledge, to embodied knowledge.

In Chapter 2, Philip Steinberg reverses the ways in which the sea is usually conceived as surface upon which people, materialities and 'landed' knowledges move (see Lambert 2005, Law 1986, Ogborn 2002), to instead think about how knowledge of the seas and oceans themselves travel. Setting sail from the Mediterranean, Steinberg explores how the ideologies bound up within an image of the Mediterranean Sea are actively and politically used to make sense of other bodies of water; the Gulf of Mexico, Caribbean Sea and Arctic Ocean, and how these bodies of water act as spaces of unity and difference. In doing so, Steinberg demonstrates how imagery of seas remains vital because it is fundamental to 'our [actual] interaction with the material sea' (Steinberg, this volume, 24).

Anne-Flore Laloë and Bärbel Bischof likewise consider the production of knowledge *about* the sea. From a historical and contemporary perspective (respectively) these authors grapple with the questions of how we 'know' the ocean. In Chapter 3, Laloë considers the role of ships' tracks in mapping the Atlantic Ocean in the nineteenth century. Here she demonstrates the difficulties in how ocean knowledge was constructed given the particular challenges of mapping a mobile space. She shows the unreliable, changeable and uncertain nature of ocean knowledge as it was produced through the ship and other technologies available for mapping the ocean. In contrast, in Chapter 4 Bischof focuses firmly on pressing contemporary concerns relating to the environmental conservation of ocean space, explicitly considering coral reef ecosystems. Bischof carefully examines the multiple and competing scientific knowledge claims which underscore how reef decline is conceptualized. In doing so, she illustrates how our knowledge of environmental degradation at sea relates directly to the ways in which we then manage water spaces. Knowledge of the ocean then, to echo Steinberg's contribution (Chapter 2) is never detached from how we actually engage with very real, physical, material water worlds.

This is exemplified in Chapter 5 by Jon Anderson, who explores the ways in which the tangible, visceral, corporeal experience of surfing on an actual, material sea creates a new knowledge both about the ocean and those who use it, and a knowledge which may extended beyond the water world to other spaces and places. Indeed, Anderson examines how the sea is not simply a static surface we move on – a point of connection between places – but rather, how the sea itself is a place; one which we are enveloped 'in' and converge 'with'. In doing so this chapter alerts us to the distinctive embodied knowledges the sea reveals, urging us to 'know' the ocean not as Lewis Carroll's 'absolute and perfect' blank, but as a lively assemblage of discrete parts; water, surfer, board, stoke; which coalesce, creating the motionful medium of the sea, as a meaningful place.

In Part II, attention is turned more explicitly to embodied engagements with the sea and the raft of corporeal experiences which are bound up with 'watery' spaces. Here the authors adopt varying approaches in attending to the ways in which ocean spaces are engaged with: as spaces we may be on (Vannini and Taggart), in (Anderson) or under (Merchant) through technologies of transportation and submersion (the car, the kayak, the wetsuit): which each result in specific visceral sensations. Building on Part I, the authors explore how such engagements lead us to reconsider and reconfigure the oceans in ways which depart from our commonplace understandings of 'watery' environments. Indeed, these chapters show how water is an unusual medium of engagement which opens up new embodied experiences compared to those possible on land. Indeed, due to water's flexible and changeable composition, each different *particular* formulation of water makes possible differing embodied engagements.

In Chapter 6 Phillip Vannini and Jonathan Taggart illustrate this by taking the reader on the open (ice) road. Through so doing they demonstrate how, in a solid elemental state, the boundary between sea and land is complicated, as it possible to literally drive across the ocean. The authors subvert and invert the engrained and take-for-grantedness of seas and oceans as liquids, thinking seriously about the very material composition of water. In Chapter 7, Anderson continues in this vein by capsizing typical written accounts of the oceans by writing of the sea *from the sea*; rather than of the sea from the land. Positioned in a sea kayak, Anderson reveals the embodied, sensual and emotional effects which arise through engagement with a motionful sea, considering the particular skill required to navigate the water, and the feelings of vulnerability, but also elation, which result. Importantly, Anderson blurs the boundaries between the bodies and water, showing how the body is not something which exists only 'on' the water's surface; but how a 'oneness' is experienced as the boundaries between nature and human life blur through the very corporeal sensations experienced by moving *through* the water. In Chapter 8, Stephanie Merchant continues to attend to the corporal sensations of experiencing water worlds through a first-hand ethnographic account of underwater diving. Focusing on the wreck of the SS *Thistlegorm* Merchant demonstrates how the oceanic experiences of divers are shaped by the very materiality of the wreck site (a materiality formed and reformed through submersion in salt water) and the memories such a site contains. Merchant also attends to the limits of the body at sea, in view of how long we can hope to be submerged underwater with the help of breathing apparatus. She thus alerts us to the adaptive practices we must embody and technological tools we must utilize to make engagements with the sea possible in the first instance.

The chapters in Part III set off by continuing to explore human life at sea, focusing on the mobile lives of fishermen in the waters adjacent to Senegal (Hallaire and Mackay) before paying attention to the more-than-human elements which are part and parcel of oceanic 'life' (Bear, Anim-Addo and Peters). Each chapter in this final part attends to the very nature of the ocean as a space filled with 'natural' elements; animals and plant life; and a space with its own more-than-human materiality which renders it a mobile and dynamic space. In particular, the chapters in this part consider, through a range of contemporary and historical case studies, the ways in which the 'nature' of the ocean influences the ways in which such a space can be regulated (or indeed, how it might elude regulation).

In Chapter 9, (following the more elite human engagements with water world presented in Part II) Juliette Hallaire and Deirdre McKay focus on fishermen in the Senegalese Atlantic, exploring how their dependency on the ocean's natural resources shapes mobile practices at sea. The authors demonstrate the fluid connections between land and sea, showing how the sea itself is a space through which landed life is sustained. They further illustrate the geopolitical tensions in governing natural oceanic resources, as both competing institutional claims to regulate ocean space ensue, and fishermen tactically seek to navigate such regulation in pursuit of their livelihoods.

Continuing to 'think about fish', in Chapter 10, Christopher Bear attends to the more-than-human geographies of the scallop fishery in Cardigan Bay, Wales. Together with Anyaa Anim-Addo (Chapter 11), more-than-human natures carried *at* sea and contained *within* the sea are the focal point of investigations, challenging anthropocentric explorations of water worlds. For Bear, thinking from the perspective of sea creatures and the seabed, allows us to reconsider how fisheries are governed, as such mobile, non-human actors and natural materialities, necessarily result in disruption and dissidence. Likewise, using the historical example of the voyages of the Royal Mail Steam Packet Company, Anim-Addo highlights how multiple factors, from yellow fever virus, to severe weather and storms, unhinged the regulation and order of shipping services, rendering them susceptible to a range of more-than-human elements. In Chapter 12, Peters continues along this tack by focusing explicitly on the sea itself as 'more-than-human'; a dynamic, vibrant matter driven by wider meteorological and extra-terrestrial forces. Focusing on the example of offshore radio broadcasting, Peters illustrates the difficulties humans face in controlling the ocean as a space with its own agencies which forever counter attempts to stabilize the sea and the experiences of those who live there. These chapters each urge us to think about power at sea (in the shape of regulatory practice), but ultimately, the power *of the* sea; as human life and trade and commerce are all challenged through the unpredictable nature of watery settings.

Conclusions

In sum, this book re-centres the oceans and seas as spaces relevant to *unearthing[1]* new understandings of the world which both move us beyond a terrestrial sphere, but also allow that terrestrial sphere to be examined in novel ways. Studying oceans and seas are essential to understanding our 'landed' lives. Water worlds cannot be conceived as 'out there' or 'irrelevant' because maritime mobilities permeate our daily existence invisibly, but significantly. That the sea *touches* our everyday lives alerts us to the material and tangible reality of water worlds. Often emptied and reduced to metaphor (Mack 2011: 25), it is vital to remember that humans do not just imagine the water world but physically experience it, and concomitantly, non-humans are not outside of the seas and oceans; they are enfolded within it in an

---

1   Meaning to move away from the terrestrial, grounded 'earth'.

embodied and enlivened way. Moreover, the seas and oceans are not merely full of people, animals and material things; they are, at the most fundamental level, constituted of matter. If we seek to bring to the fore the various ways the seas and oceans are 'filled', we can attempt to write about the world in a different way, from perspectives which do not privilege the land, or land-based thinking. Gaining novel and important insights from the water world enables us to create a new language, a 'Thalassology', for conceptualizing watery-human interactions and which may be employed at, but also beyond the oceans and seas.

References

Anderson, J. 2012. Relational places: the surfed wave as assemblage and convergence. *Environment and Planning D: Society & Space*, 30, 570–87.

Artimage, D. and Braddick, M.J. 2002. *The British Atlantic World 1500–1800*. Basingstoke: Palgrave Macmillan.

Barnes, T. and Duncan, J. 1992. *Writing Worlds: Discourse, Texts, and Metaphors in the Representation of Landscape*. London: Routledge.

Bear, C. and Eden, S. 2008. Making space for fish: the regional, networked and fluid spaces of fisheries certification. *Social and Cultural Geography*, 9(5), 487–504.

Bennett, J. 2004. The Force of Things: Steps Towards an Ecology of Matter. *Political Theory*, 32(3), 347–72.

Bourdieu, P. 1977. *Outline of a Theory of Practice*. Cambridge: Cambridge University Press.

Bourdieu, P. 1991. *Language and Symbolic Order*. Cambridge: Polity Press.

Carroll, L. 2009 edition. *The Hunting of the Snark*. Auckland: The Floating Press.

Cooney, G. 2003. Introduction. Seeing land from the sea. *World Archaeology*, 5(3), 323–8.

Craciun, A. 2010. The frozen ocean. *PMLA*, 125(3), 693–702.

Cresswell, T. 2000. Falling down. Resistance as diagnostic. In *Entanglements of Power: Geographies of Ddomination/Resistance*, edited by J. Sharp et al. London: Routledge. 260–73.

Cresswell, T. 2004. *Place: A Short Introduction*. Oxford: Blackwell Publishing.

Cresswell, T. 2006. *On the Move: Mobility in the Modern Western World*. London: Routledge.

Davidson, J., Bondi, L. and Smith, M. 2005. *Emotional Geographies*. Farnham: Ashgate.

DeLanda, M. 2006. *A New Philosophy of Society: Assemblage Theory and Social Complexity*. London: Continuum.

Deleuze, G. and Guattari, F. 2004. *A Thousand Plateaus*. London: Athlone Press.

Dovey, K. 2010. *Becoming Places: Urbanism/Architecture/Identity/Power*. Abingdon: Routledge.

Featherstone, D. 2005. Atlantic networks, antagonisms and the formation of subaltern political identities. *Social and Cultural Geography*, 6(3), 387–404.

Gilroy, P. 1993. *The Black Atlantic: Modernity and Double Consciousness.* London: Verso.

Harvey, D. 1996. *Justice, Nature, and the Geography of Difference.* Oxford: Blackwell.

Hill, L. Abbott, J. 2009. Surfacing tension: Toward a political ecological critique of surfing representations. *Geography Compass*, 3, 275–96

Ingold, T. 2008. Bindings against boundaries: Entanglements of life in an open world. *Environment and Planning A*, 40, 1796–810.

Jackson, S.E. 1995. The water is not empty: Cross-cultural issues in conceptualising sea space. *Australian Geographer*, 26(1), 87–96.

Jones, O. 2011. Lunar-solar rhythmpatterns: Towards the material cultures of tides. *Environment and Planning A*, 43, 2285–303.

Lambert, D. 2005. The counter-revolutionary Atlantic: White West Indian petitions and proslavery networks. *Social and Cultural Geography*, 6(3), 405–20.

Lambert, D., Martins, L. and Ogborn, M. 2006. Currents, visions and voyages: Historical geographies of the sea. *Journal of Historical Geography*, 32(3), 479–93.

Langewiesche, W. 2004. *The Outlaw Sea: A World of Freedom, Chaos, and Crime.* Granta London: Books.

Lavery, B. 2005. *Ship: 5000 Years of Maritime Adventure.* UK: Dorling Kindersley Publishers Ltd.

Law, J. 1986. On methods of long-distance control: Vessels, navigation and the Portuguese route to India. In *Power, Action and Belief: A New Sociology of Knowledge?*, edited by J. Law. London: Routledge. 234–63.

Levi-Strauss, C. 1973. *Tristes Tropiques.* London: Jonathan Cape Ltd.

Lorimer, H. 2005. The busyness of 'more-than-representational.' *Progress in Human Geography*, 29, 83–94.

Mack J, 2011. *The Sea: A Cultural History.* London: Reaktion.

Malinowski, B. 1922. *Argonauts of the Western Pacific: An Account of Native Enterprise and Adventure in the Archipelagos of Melanesian New Guinea.* London: Routledge & Sons.

Ogborn, M. 2002. Writing travels: Power, knowledge and ritual on the English East India Company's early voyages. *Transactions of the Institute of British Geographers*, 27(2), 155–71.

Peters, K. 2010. Future promises for contemporary social and cultural geographies of the sea. *Geography Compass*, 4(9), 1260–72.

Peters, K. 2012. Manipulating material hydro-worlds: Rethinking human and more-than-human relationality through off-shore radio piracy. *Environment and Planning A*, 44, 1241–54.

Philbrick, N. 2001 *In the Heart of the Sea.* London: Harper Collins.

Pile, S. 2010. Emotions and affect in recent human geography. *Transactions of the Institute of British Geographers*, 35, 5–20.

Raban, J. 1999. *Passage to Juneau: A Sea and its Meanings.* Basingstoke and Oxford: Picador.

Shields, R. 1992. *Places on the Margins*: *Alternative Geographies of Modernity.* London: Routledge.

Steinberg, P.E. 1999a. Navigating to multiple horizons: Towards a geography of ocean space. *Professional Geographer*, 51(3), 366–75.

Steinberg, P.E. 1999b. The maritime mystique: Sustainable development, capital mobility, and nostalgia in the world-ocean. *Environment and Planning D: Society & Space*, 17, 403–26.

Steinberg, P.E. 2001. *The Social Construction of the Ocean.* Cambridge: Cambridge University Press.

Steinberg, P.E. 2013. Of other seas: metaphors and materialities in maritime regions. *Atlantic Studies*, 10(2), 156–69.

Tally, R.T. 2013. *On Literary Cartography: Narrative as a Spatially Symbolic Act* [Online] Available at: http://www.nanocrit.com/essay-two-issue-1-1/ [accessed 1 March 2013].

Thrift, N. 2004. Intensities of feeling: Towards a spatial politics of affect. *Geografiska Annaler B*, 86 (1), 57–8.

Tuan, Y-F. 1991. Language and the making of place: A narrative-description approach. *Annals of the Association of American Geographers*, 81(4), 684–96.

Westerdahl, C. 2005. Seal on land, elk at sea: Notes on and applications of the ritual landscape at the seaboard. *The International Journal of Nautical Archaeology*, 34(1), 2–23

Whatmore, S. 2006. Materialist returns: Practising cultural geography in and for a more-than-human world. *Cultural Geographies*, 13, 600–609.

Wood, N. Smith, S. 2004. Instrumental routes to emotional geographies. *Social and Cultural Geography*, 5 (4), 533–48.

Young, M. 1979. *The Ethnography of Malinowski: The Trobriand Islands 1915–18.* London: Taylor Francis.

PART I
Ocean Knowledges:
Understanding the Water World

Chapter 2

Mediterranean Metaphors: Travel, Translation and Oceanic Imaginaries in the 'New Mediterraneans' of the Arctic Ocean, the Gulf of Mexico and the Caribbean

Philip E. Steinberg

Oceans are 'known' in many different ways. As the chapters of this book reveal, scientists, sailors, surfers, passengers and divers all have their own perspectives on the ocean as a fluvial, dynamic arena of human and non-human biota, of minerals and molecules, of affects and ideologies.

But what of the perspective from *beyond* the sea? Does that even need to be considered in this book? After all, this is a book that emphasizes *affect* and *experience* in a resolutely *material* sea. Do we need to provide a forum here for those who view the ocean simply as a surface in the middle – whether as a space to be crossed, or plundered, or ignored, or as a space that merely divides or connects?

I answer this with an emphatic 'yes.' For better or worse, our perceptions of the ocean are structured not just by the tactile *experiences* that we have with its liquid element but by the *stories* that we tell about the sea, including the simplified stories of functionality or the non-stories of absence. Narrated understandings, even if not derived from sensory experience, contribute to the ocean assemblage. Indeed, as the paradigmatic space of the sublime – where emotional understanding exists on a plane removed from cognition – the ocean derives much of its power from the reproduction of its image, including by those who never come in contact with, or sail across, its waters. Like a map (Del Casino and Hannah 2006, Kitchin and Dodge 2007), an ocean is more-than-representational. It is continually reconstructed through our encounters, but as we engage the sea our experiences are performed and internalized through articulations with pre-existing imaginaries.

To be clear, this call for taking imagined oceans seriously should not be seen as an endorsement of a perspective wherein the ocean is reduced to a metaphor – a signifier for cultural hybridity or global commerce or any of the other social processes that the ocean has been made to stand for in recent work in literary, cultural and historical studies. Indeed, elsewhere I specifically reject this perspective (Steinberg 2012, see also Blum 2010). But images of the ocean *do* matter, not because they exist apart from, or after, our interaction with the material sea but because they contribute to that interaction and, thereby, to its social (and more-than-social) construction. To that end, this chapter focuses on one specific imagined ocean – the Mediterranean – and how its image has been applied to construct meanings and practices in other maritime regions.

The Mediterranean as an Ocean of Connection and Division

Oceans have long been seen in Western thought as barriers. In Macrobius' worldview the oceans were impassable. A world was believed to exist on the other side of the ocean, but it was inaccessible because it lay across the torrid zone. In the medieval world of the mappamundi, the ocean had even less potential as a surface for connection; it was simply a limit, and no earthly space of note, certainly nothing that was mappable, was believed to exist on the other side (Cosgrove 2001, Edson 1997, Gillis 2004, Harley and Woodward 1987). In the modern era, the world typically has been characterized as a universe of continents, wherein the oceans that exist between these fundamental land masses serve simply to divide the terrestrial spaces that matter (Lewis and Wigen 1997). From this continental perspective, the ocean and its uses occur outside the space of society, subsequent to the constitution of state territories and in defiance of their terrestrial roots (Steinberg 2001).

Notwithstanding the predominance of this worldview, for at least a century it has been challenged by academic observers from geography and beyond. Early in the twentieth century, Ellen Churchill Semple (1911: 294) noted that the ocean on which 'man explores and colonizes and trades' is no less a space of society than the land on which 'he plants and builds and sleeps.' At mid-century, Richard Hartshorne (1953: 386) addressed the continentalist perspective directly, stating that although '[oceans] do divide, they do not separate.' Since that time, scholarly communities have arisen to study a number of ocean basins (see Lewis and Wigen 1999), and although these communities vary with respect to disciplinary or interdisciplinary focus they all share a perspective in which the ocean is moved from the margin to the centre of the regional social formation.

Like any antithetical categorization, the ocean region can either challenge or reproduce the fundamental assumptions of the dominant construction to which it is posed as an alternative. On the one hand, when one designates an ocean as the element that unites a region, fluidity and connections replace embeddedness in static points and bounded territories as the fundamental nexus of society and space. 'Roots' are replaced by 'routes', and this suggests a radical ontology of deterritorialization and reterritorialization (cf. Deleuze and Guattari 1987). On the other hand, by reaffirming the concept of the region as a unit of analysis – a unit that is stable in space and time and, therefore, potentially explanatory – the ocean region perspective can inadvertently reproduce the static and essentialist spatial ontology that it attempts to subvert.[1]

Giaccaria and Minca (2011) advance this critique by identifying ocean basin-based regions as exemplarily postcolonial spaces that reproduce and naturalize ideals of unity in difference. On the one hand, the ocean in the middle of a maritime region links spaces and societies that are purported to be 'naturally' different. The different societies exist on opposite sides of a seemingly natural divide, a purportedly empty and separating ocean. On the other hand, because the ocean connects, even if it does not homogenize, the societies in an ocean region appear to exist in a permanent and natural universe of exchange and interaction that reproduces difference. Existing within an idealized arena of connectivity amidst difference, the various societies within an ocean region are linked together in an arena of mobility in which all entities – those with relatively more power and those with relatively less – are transformed even as they resist the 'other'.

While all ocean regions are, in this sense, prefiguratively postcolonial, arguably the paradigmatic case is the Mediterranean (Chambers 2008). In part, this is because of the Mediterranean's physical geography (relatively small and enclosed), in part it is because of its location at the intersection of Europe and one of its longest standing 'others' (the Arab 'orient'), and in part it is because of the long history in the humanities of treating the Mediterranean as a singularly unified, but also resolutely divided, region (Giaccaria and Minca 2011). For all these reasons,

> [amidst] a paradoxical interplay between different (and potentially conflictual) representations of this sea that alternate narratives of homogeneity and continuity with those of heterogeneity and discontinuity, … [the rhetoric of mediterraneanism sustains] the belief in the existence of *a geographical object called the Mediterranean*, where different forms of proximity (morphological, climatic, cultural, religious, etc.) justify a specific rhetorical apparatus through the production of a simplified field of inquiry, otherwise irreducible to a single image. (Giaccaria and Minca 2011: 348, emphasis in original)

The Mediterranean thus comes to be seen as something that, although permanently divided, is also permanent in its wholeness: 'The *mediterraneisme de la fracture* [is understood as] … something substantially immutable – a vision that resembles, in many ways, the cultural "containers" imagined and celebrated in Orientalist colonial rhetoric and Romantic literature' (Giaccaria and Minca 2011: 353).

In the remainder of their article, Giaccaria and Minca discuss ways in which one can harness the alterity that lies within the mediterraneanist discourse without

---

1   For a broader discussion of how revisionist approaches to the region can inadvertently reproduce the ontologies that they seek to undermine, see Smith (1987) and Marston (2000). For expansions of this discussion that refer specifically to ocean-basin regions, see Giaccaria and Minca (2011) and Steinberg (2012).

inadvertently reproducing its orientalism. Even as they pursue this agenda, however, they fail to remark on how the ultimate power of mediterraneanism, like all forms of orientalism, lies not simply in reproducing an ideal of stabilized difference but in the designation of this unity as a *category* that can then be integrated into systems of language and meaning that, in turn, are used to 'understand' (and thereby design futures for) other peoples (Mignolo 2003, 2005). Put another way, the power of mediterraneanism (the idea of there being a distinct Mediterranean region) flows not just from its purported ability to explain the (upper-case) Mediterranean as a naturalized arena of linked difference but from its functionality as a *category* wherein the presence of an inner sea (a lower-case mediterranean) is used to explain a generalized condition of difference amidst connection. In short, the power of the Mediterranean idea derives not just from its representation of a divided but unified ocean basin as 'something substantially immutable' but from the idea's existence as an 'immutable mobile' (Latour 1987), an idea that travels.

When an idea travels, however, it is never truly immutable. Although ideas gain their power through travel, travel invariably necessitates translation, and translation will always modify the power of an idea, even as it provides the means for realizing (or performing) that power (Clifford 1997). As is shown in this chapter, the mediterraneanist ideal of an inner sea that simultaneously essentializes both unity and difference has had specific histories as it has travelled to other regions. Below, I relate two stories of mediterraneanist travel and translation: first in the Gulf of Mexico/Caribbean, and then in the Arctic.

The Travelling Mediterranean 1: The Gulf of Mexico and the Caribbean

If mediterraneanism achieves its power by constructing the idea of the inner sea as a space that facilitates both fractures (or difference) and crossings (or unity), then at first glance it appears as if only the first of these is present in the Gulf of Mexico and the Caribbean. From the perspective of the United States, the Gulf of Mexico consists of a series of fragmented coastal destinations: a peninsular Florida coast of pristine beaches and recreational fishing, a Texas/Louisiana coast of off-shore oil drilling and shrimping communities, and, in the middle, a Mississippi/Alabama/ Florida panhandle coast of shipyards, naval bases, and casinos. Further to the east (and south), the Caribbean is understood as a series of island isolates, what Epeli Hau'ofa (1993) calls 'islands in a sea' in contrast with the integrated South Pacific 'sea of islands' that he extols.

While each of these Gulf/Caribbean images is certainly maritime, none of them suggests an underlying historical, or ongoing, space of maritime unity – a mediterranean space of crossings. Mexico, which might logically be perceived as lying on the 'other' side of the region (the equivalent of North Africa and the Levant, in the Mediterranean context), is instead seen as an extension of the arid western United States, not a space that is joined to the United States through maritime connectivity. This geographic erasure in U.S. thought, in which the southern maritime frontier is subsumed by the western land frontier, is reproduced in the Hollywood Western, where Mexico is almost universally depicted as an extension of the southwestern U.S. desert, not the land that lies across from the Gulf coast of the southeastern United States. The resulting conception of the Gulf region as a series of local destinations, as opposed to being an integrated maritime space unified by a body of water, is so pervasive that when Mississippi state legislator Steve Holland proposed renaming the Gulf of Mexico the Gulf of America in an effort to spoof his anti-immigration colleagues the joke was lost on the national media (Wilkinson 2012).[2]

However, this construction of the Gulf/Caribbean as an ocean without routes of connection is relatively new; it is no more 'natural' or static than the prevalent conception of the Mediterranean as a sea that joins connectivity with difference. During the first decades of the twentieth century, when the United States was asserting its regional ambition in the wake of the Spanish-American War and the newly opened Panama Canal, the U.S. perspective on the region was much more mediterraneanist. Indeed, analogies to the Mediterranean were explicitly deployed to signify the region's potential as an arena in which the United States could expand its frontiers through maritime connectivity to different, but accessible, places. For instance, at the turn of the twentieth century, when the American naturalist and travel writer Frederick Ober (1904) published *Our West Indian Neighbors*, he subtitled the book *The Islands of the Caribbean Sea, "America's Mediterranean", their Picturesque Practices, Fascinating History and Attractions for the Traveler, Nature Lover, Settler, and Pleasure Seeker.* 1920s tourism promotion material for Key West, Florida (at Florida's southern tip) and the Florida East Coast Railway (which ran trains to Key West) similarly identified the Caribbean not as the *end* of America but as its continuation.[3] Key West itself was promoted as the 'Gateway to Cuba, West Indies, Panama Canal, Central America, and South America,' while on promotional maps from this era railroad lines merged seamlessly into ferry routes, and ocean shallows and reefs were coloured so that one had to concentrate to ascertain where land ended and water began. Most suggestively, Key West was labelled 'America's Gibraltar,' a geographic appellation that, even more than that of the Mediterranean as a whole, resonates with dual connotations of connection and difference.[4]

Today, not only have these Mediterranean signifiers disappeared, but even the less specific mediterraneanist images of the Gulf/Caribbean as an inner sea of connection amidst difference are absent. On Key West, major tourist icons such

---

2   The gaps in the U.S. imagination of the Gulf of Mexico and the Caribbean as a maritime region are explored further by Steinberg (2011) and Silva Gruesz (2006).

3   This trope of ocean basins expanding the U.S. frontier was reproduced later in the century in the Pacific, although not typically with direct reference to the Mediterranean (Connery 1994).

4   The Key West historical promotional materials referred to in this paragraph are available for viewing at the Flagler Station Over-Sea Railway Historeum in Key West (http://www.flaglerstation.net).

as the 'Southernmost Point' monument (marking the southernmost point of the continental United States) and the 'Mile 0' marker (signifying the terminus of U.S. Highway 1, which runs the length of the country's east coast) suggests that there is nothing to connect to beyond Florida's shores. Indeed, in some senses the contemporary U.S. perspective on the Gulf of Mexico and the Caribbean resembles that of a mappamundi: The ocean appears simply as a limit, an end beyond which there is no known civilization (or even potential for civilization). Nothing, or certainly nothing of interest, exists beyond the coastal zone. In today's de-mediterraneanized Gulf/Caribbean it is impossible to conceive of the ocean as the central binding element of a diverse region constituted by the various societies along its shores. Instead, the ocean is reduced to constituting the outer boundary of a series of coastal and island destinations.[5]

The Travelling Mediterranean 2: The Arctic[6]

In the 1920s, just as the Mediterranean metaphor was being deployed in the Caribbean and the Gulf of Mexico to justify U.S. commercial (and military) expansion, a similar trope was being applied in the Arctic. A key proponent of this effort was the Canadian-American anthropologist Vilhjalmur Stefansson. In such books as *The Friendly Arctic* (1921) and *The Northward Course of Empire* (1922a), Stefansson argued that the Arctic, far from being a frozen wasteland, was a 'Polar Mediterranean.' Like the Mediterranean, he wrote, the Arctic featured a relatively navigable central space that united diverse coastal peoples in commerce and productive interaction. With the advent of new transportation technologies (especially the airplane) that would further ease navigation across its inner sea, the Arctic was likely to emerge as a new centre of civilization. As Stefansson wrote in *National Geographic Magazine*:

> A map giving one view of the northern half of the northern world shows that the so-called Arctic Ocean is really a Mediterranean sea like those which separate Europe from Africa or North America from South America. Because of its smallness, we would do well to go back to an Elizabethan custom and call it not

5   Further reflections on the ways in which residents and tourists in Key West construct and deny connections with lands and waters to the south (as well as with the mainland United States to the north) can be found in Steinberg (2007) and Steinberg and Chapman (2009).

6   Interviews discussed in this section were conducted in 2010 as part of a U.S. National Science Foundation-funded project on Territorial Imaginaries and Arctic Sovereignty Claims. The interviews cited here were conducted by myself, Mauro Caraccioli, and Jeremy Tasch (Anchorage); myself, Mauro Caraccioli, Jeremy Tasch, and Elizabeth Nyman (Washington); and Hannes Gerhardt (Nuuk and Oslo). I am grateful to each of these colleagues for their assistance in gathering and sharing interview data.

the Arctic Ocean but the Polar Sea or Polar Mediterranean. The map shows that most of the land in the world is in the Northern Hemisphere, that the Polar Sea is like a hub from which the continents radiate like the spokes of a wheel. The white patch shows that the part of the Polar Sea never yet navigated by ships is small when compared to the surrounding land masses. (Stefansson 1922b: 205)

Although the Mediterranean analogy has faded from public imagination in the Caribbean and the Gulf of Mexico, it has recently undergone a resurgence in the Arctic. Regional boosters are seeking new tropes to assist them in realizing the opportunities that they believe will come from climate change and the demise of Cold War tensions, and they have seized upon the Mediterranean as an appropriate, if climatically incongruous, analogue.

The prevalence of mediterraneanist discourse in the Arctic became apparent to my research team in 2010 when we were in Alaska conducting interviews on Arctic geopolitics. Driving to an interview on our first day in Anchorage we turned on Alaska Public Radio and heard a promotional advertisement announcing that the topic for that week's Talk of Alaska radio show was to be: 'Is the Arctic the new Mediterranean?' As it turned out, the programme featured an interview with author Charles Emmerson (who had recently published *The Future History of the Arctic* [2010]). The programme began with the host, Steve Heimel, giving the following introduction:

The climate is warming … and what this does is it opens up the Arctic Ocean. The question is: What now? Does the Arctic at some point become the Mediterranean? Does the Arctic Ocean become the centre of civilization, with a warming climate, and how many years from now does what happen? Everybody's looking at this. Anybody in the policy area, whether it's science, whether it's mineral exploration, whether it's international law. The implications of an opening Arctic are staggering and different from the kinds of dreams that the polar explorers once had, but not all that different.

Emmerson, in turn, followed up by discussing the predictions of one of those 'polar explorers':

Vilhjalmur Stefansson … wrote this wonderful book called *The Northward Course of Empire*, and his idea was that thousands of years ago the centres of civilization had all been in very hot places, very temperate places – Cairo, Baghdad, then in Rome and later slightly cooler places such as London and Paris and then New York – his vision was that that line is going to continue and the centre of civilization 20 or 30 years from now is going to be Winnipeg or Anchorage. (transcribed from Alaska Public Radio Network 2010)

Apparently, the image of an Arctic Mediterranean resonated with Alaskans, because it was brought up, without our prompting, in two subsequent interviews.

First, an official from Alaska's Department of Natural Resources specifically referenced the radio programme:

> Last Tuesday there was a public radio programme produced here by the Alaska Public Radio Network called Talk of Alaska and last week was 'Is the Arctic the New Mediterranean?' ... I didn't make it – I can't listen to it during the day – but I thought that the tag was sort of funny. I'd never thought of the Arctic as a new Mediterranean, but there's probably no [fewer] people living around the Arctic [now] than lived around the Mediterranean in Roman times.

In another interview, an elected official, when informed that we were a team of geographers funded by the U.S. National Science Foundation, remarked:

> The first geography teacher I ever had suggested that civilizations come about because of interaction and accessibility, and, you know, that's why classic geography says, 'Look at rivers, ports, how does that all happen?' ... [With climate change,] I'd be fascinated to kind of take the classical geographic thinking and say, 'Okay, what will be the ties that will endure, that will ultimately change this?' And I don't know what they are; there's a lot of potential right now and there are some things that [will] work and things that won't work, but now, you take a look at the vision of the Arctic, where the unified Arctic is kind of a Mediterranean play and so forth .... And that is the question that I'd ask if I had a National Science Foundation grant to look at cooperation in the Arctic. Because in the end, all the meetings we do, they're fun meetings, but what endures is what people are going to do with Arctic regular commerce.

In making this statement, the elected official was unknowingly (or, perhaps, knowingly) referencing a mediterraneanist, liberal view of the ocean held since Aristotle, who encouraged civilizations and cities to adopt a maritime orientation in the belief that cultural and commercial movement across the maritime surface of exchange would lead to peace and prosperity (Gottman 1973).

Although not directly referencing the Mediterranean, then Soviet Premier Mikhail Gorbachev similarly drew on the mediterraneanist concept of the cosmopolitan *emporion* when he noted, in 1987, 'The Arctic is not only the Arctic Ocean but also ... the place where the Eurasian, North American, and Asia pacific regions meet, where the frontiers come close to one another and the interests of states ... cross' (Gorbachev 1987). Likewise, then U.S. vice presidential candidate Sarah Palin famously proclaimed, 'We have that very narrow maritime border between the United States ... and Russia .... They're very, very important to us and they are our next door neighbor .... You can actually see Russia from land here in Alaska, from an island in Alaska .... I'm giving you that perspective of how small our world is and how important it is that we work with our allies, to keep good relation[s] with all of these countries, especially Russia' (ABC News 2009).

As in the actual Mediterranean, however, the mediterraneanist ideal of unity amidst exchange in the Arctic is paired with one of antagonistic difference. Resurrecting the dream of Vilhjalmur Stefansson, military affairs journalist Barry Zellen (2008) wrote in Toronto's *Globe and Mail* that the only factor that had so far prevented the Arctic from emerging as a site of interaction and investment was its climate. Now, with climate change, he argued, the potential for the Arctic to emerge as a 'new Mediterranean' will likely be realized. However, Zellen went on to write that the 'Age of the Arctic' that will emerge is as likely to be one of hostile interaction among antagonists as one of peaceful commerce. Giving a nod to Sir Halford Mackinder, Zellen concluded by warning that a proactive military strategy is warranted because 'the long isolated "Lenaland" along the Arctic basin will transform into a highly productive and strategically important "Rimland," transforming the Arctic into tomorrow's equivalent of the Mediterranean, a true strategic, economic and military crossroads of the world.' In a similar vein, former U.S. Coast Guard officer Scott Borgerson (2008) has likened the melting Arctic to the Mediterranean as a parallel zone of contestation. Borgerson argues that the Mediterranean has avoided outright conflict because antagonistic nations on opposite sides of its connecting waters have recognized each other's coastal rights and the rule of law, and he argues for reaching a similar consensus in the Arctic before it is too late.

Gorbachev, Palin, Zellen, and Borgerson all deploy the physical imagery of an inland sea to construct the Arctic as a region, like the Mediterranean, that combines unity with division in Giaccaria and Minca's (2011: 348) 'paradoxical interplay ... that alternate[s] narratives of homogeneity and continuity with those of heterogeneity and discontinuity.' However, the flexibility of the mediterranean image – in which the fluidity of the sea both erases and magnifies difference – and the intensity of the 'paradoxical interplay' that results allows individuals to use the image to support very different political diagnoses. For Gorbachev and Palin, the paradoxical qualities of the mediterranean Arctic are used to highlight its potential as a space of peace wherein differences may be overcome through commerce and exchange. For Zellen and Borgerson, these same qualities are used to highlight its potential as a space of conflict wherein encounters between naturally separated nations are likely to breed distrust and acrimony.

This theme of the Arctic as a Mediterranean-like meeting place where the West brushes up against its 'others' was echoed in an interview that we conducted with a U.S. State Department official who effectively combined the Zellen/Borgerson position (as a divided space, the Arctic is a natural space of discord) with the Gorbachev/Palin position (that maritime space of division is also a space of connection, and thus it provides an environment in which this discord may be overcome):

A difficulty for the Russians is learning how to be part of the community. The old Soviet ways of doing things still seems to be ingrained in them .... I think part of it is the Russian mindset, and getting past that .... In some ways, it's like

[the Russians] have to learn to play nice with others, and the Arctic may be the
place to do that.

For this official, the fluid centre of the Arctic, like the Mediterranean, creates
divisions (in particular, it makes it possible for proximate neighbours to have
different social skills in the international community) but it also creates the
connections that could lead to transcendence of those divisions. Interaction between
these disparate neighbours is seen as potentially productive and increasingly likely
due to connections brought about by an inland sea. However, such connections
are also recognized as inevitably fraught with tension due to superorganic cultural
differences that result from the geographic division mediated by that same
intervening ocean.[7]

Still others have used comparisons with the Mediterranean to argue for
heightened national presence in the Arctic region. For instance, an official from the
Government of Greenland likened the Arctic to the Mediterranean not as a zone of
interaction but as a region whose borders ended at some point 'inland.' Thus, he
argued that just as non-Mediterranean countries with non-territorial interests in the
region should have, at most, observer status in Mediterranean governance issues,
non-Arctic countries should have little power in governing the Arctic.[8] In similar
fashion, in 2010 the Standing Committee on National Defence in Canada's House
of Commons issued a report on 'Canada's Arctic Sovereignty' (Government of
Canada 2010) that began with a discussion of Stefansson's predictions about the
future centrality of the Arctic and then went on to note that in the context of climate
change and the end of the Cold War Canada must assert its Arctic sovereignty to
foster national development in its portion of Stefansson's rapidly emergent Polar
Mediterranean.

Conclusion

As a response to the continentalism that pervades our understanding of the world,
numerous scholars and commentators have proposed that we conceive of a world
of mediterranean regions. These regions' oceanic centres allow us to highlight such

---

7   Bravo (2009) discusses more generally how Russia is constructed as an 'other' in
the literature on policy responses to climate change in the Arctic region.

8   Interestingly, a non-Arctic member of the European Union's delegation to the
Standing Committee of Parliamentarians of the Arctic Region constructed a parallel
between the Arctic and the Mediterranean to make just the opposite case, arguing *for* the
inclusion of non-Arctic countries in Arctic decision making: 'The EU is partly Arctic in a
way and I mean, you can also compare it to the Mediterranean policy of the EU. I mean,
not all EU member states are Mediterranean countries obviously. Maybe it is more obvious
that the EU should have an interest in the Mediterranean region, but I mean you could also
use the same argument and say that all the European states are Mediterranean states.'

phenomena as intensified economic exchange (e.g. Dirlik's [1998] 'Pacific Rim'), transnational diasporas (e.g. Gilroy's [1993] 'Black Atlantic'), or a 'planetarity' in which individuals are both local and global (e.g. Craciun's [2009] circumpolar Arctic). In all of these visions, the character of the ocean at the centre of the region as a space that simultaneously facilitates movement and divides plays a crucial role in establishing norms of connectivity amidst difference. As Giaccaria and Minca (2011) and Chambers (2008) note, this narrative has a long history in the Mediterranean and, as I have argued here, the mobility and transferability of these ideas of mediterraneaness have facilitated the Western understanding of the ocean region as a postcolonial space.

Two problems emerge, however, when one looks at the world as a series of mediterraneans. The first, which has barely been touched on in this chapter but which should be evident from the other chapters in this book, is that, to quote Hester Blum (2010: 670), 'the sea is not a metaphor.' So long as one views the ocean in the abstract as a central space – even as a central space that performs a complex mix of functions that serve to both unite and divide – one runs the risk of forgetting that the sea is constructed not just through how we *think* about it but through how we *experience* it, as a material space that is encountered by our embodied practices as well as the practices of the ocean's other (human and non-human) constitutive elements. Any perspective on an ocean region that focuses on terrestrial societies that are simply 'linked by' or 'divided by' (or 'linked *and* divided by') a central sea are necessarily incomplete.

The second, related problem is that once the ocean in the centre is reduced to a metaphor then the ocean region similarly can become a metaphor, and, as scholars of metaphor caution, the application of a metaphor, particularly to a space, is never innocent (Brown 2000, Smith and Katz 1993). In each of the examples from the Caribbean, the Gulf of Mexico, and the Arctic, the application of the 'Mediterranean' (a specific place) to a 'mediterranean' (a region characterized by a central sea) is partial. Those who apply signifiers of mediterraneanism desire some aspects of the Mediterranean idea to travel (e.g. the idea of 'civilization' spreading to 'opposite' shores) while disabling others (e.g. the potential for migration in the reverse direction). As just one example, interviews conducted in Key West revealed that tourists and residents there had contradictory attitudes toward a hypothetical bridge to Cuba: they supported the benefits that a bridge would bring for tourist mobility and the southward diffusion of cultural norms (in this case, specifically, tolerance of homosexuality) while fearing that a bridge would facilitate an influx of Cuban migrants (Steinberg 2007, Steinberg and Chapman 2009).

Just as the Mediterranean Sea is more than an imagined surface, the Mediterranean region is more than a trope for understanding (or performing) postcolonial dynamics of connection amidst division. Metaphorical references to the Mediterranean do indeed have these affective resonances, in part because of the properties attributed to the ocean in the Western imagination and in part because of the specific Western history of thinking both about and with the Mediterranean. But the application of mediterraneanism as a category – its travel as a purportedly immutable mobile – inevitably involves translation, and this translation occurs in specific historic contexts. The very fact that Mediterranean references in the Gulf of Mexico and the Caribbean have faded into obscurity while they are on the rise in the Arctic suggests that, despite its appeal to historic constancy and 'natural' physical geography, the mediterraneanization of a region is dependent as much on the ambitions of those who would deploy the trope as on the actual dynamics of connectivity and difference that are present in the region. In particular, as a concept that justifies sustained interaction among nations that are deemed to be at different levels of civilizational development, mediterraneanization and the fetishization of the ocean as a space that connects while preserving heterogeneity appears most prominently in times and spaces of imperial expansion. Many imperialist ideologies join the idealization of natural difference between societies with the belief that integration of one society into the other is possible, and inland seas are well positioned to aid in the pairing of these seemingly contradictory narratives.

The ocean in the centre of the Mediterranean 'water world' thus cannot be reduced to a surface, whether a surface of division, a surface of connection, or – as in the case of the postcolonial Mediterranean sea – a surface of division amidst connection. It is all of these, and the image of it as all of these is regularly applied to construct and interpret a universe of mediterranean 'water worlds.' But this application of the image occurs amidst actual practices by which connection and division are produced, both in and around the water. The representation of the inner sea as postcolonial space is *one* component of a wet ontology, but to fully understand the 'water worlds' that constitute our planet this imagery must be joined with the historically and geographically specific material practices – of humans and others – that constitute marine space.

References

ABC News. 2009. Full transcript: Charlie Gibson interviews GOP Vice Presidential candidate Sarah Palin [Online, 23 November] Available at: http://abcnews.go.com/Politics/Vote2008/full-transcript-gibson-interviews-sarah-palin/story?id=9159105&page=4 [accessed: 10 June 2012].

Alaska Public Radio Network. 2010. Talk of Alaska: The future of the Arctic [Online, 17 June] Available at: http://www.alaskapublic.org/2010/06/17/talk-of-alaska-the-future-of-the-arctic/ [accessed: 10 June 2012].

Blum, H. 2010. The prospect of oceanic studies. *Proceedings of the Modern Language Association*, 125(3), 670–77.

Borgerson, S.G. 2008. Arctic meltdown: The economic and security implications of global warming. *Foreign Affairs*, 87(2), 63–7.

Bravo, M.T. 2009. Voices from the sea ice: The reception of climate impact narratives. *Journal of Historical Geography*, 35(2), 256–78.

Brown, M.P. 2000. *Closet Space: Geographies of Metaphor from the Body to the Globe*. London: Routledge.

Chambers, I. 2008. *Mediterranean Crossings: The Politics of Interrupted Modernity.* Durham, NC: Duke University Press.

Clifford, J. 1997. *Routes: Travel and Translation in the Late Twentieth Century.* Cambridge, MA: Harvard University Press.

Connery, C.L. 1994. Pacific Rim discourse: The U.S. global imaginary in the late Cold War years. *Boundary 2*, 21(1), 30–56.

Cosgrove, D. 2001. *Apollo's Eye: A Cartographic Genealogy of the Earth in the Western Imagination.* Baltimore: Johns Hopkins University Press.

Craciun, A. 2009. The scramble for the Arctic. *Interventions*, 11(1), 103–14.

Del Casino, V.J. Jr. and Hanna, S.P. 2006. Beyond the 'binaries': A methodological intervention for interrogating maps as representational practices. *ACME: An International E-Journal for Critical Geographies*, 4, 34–56.

Deleuze, G. and F. Guattari. 1987. *A Thousand Plateaus: Capitalism and Schizophrenia.* Minneapolis: University of Minnesota Press.

Dirlik, A. (ed.). 1998. *What Is In a Rim? Critical Perspectives on the Pacific Region Idea, 2nd edition.* Lanham, MD: Rowman & Littlefield.

Edson, E. 1997. *Mapping Time and Space: How Medieval Mapmakers Viewed their World.* London: British Library.

Emmerson, C. 2010. *The Future History of the Arctic.* Washington, DC: Public Affairs.

Giaccaria, P. and Minca, C. 2011. The Mediterranean alternative. *Progress in Human Geography*, 35(3), 345–65.

Gillis, J. 2004. *Islands of the Mind: How the Human Imagination Created the Atlantic World.* New York: Palgrave Macmillan.

Gilroy, P. 1993. *The Black Atlantic: Modernity and Double Consciousness.* Cambridge, MA: Harvard University Press.

Gorbachev, M. 1987. Mikhail Gorbachev's speech in Murmansk at the ceremonial meeting on the occasion of the presentation of the Order of Lenin and the Gold Star to the City of Murmansk, October 1, 1987 [Online, 1 October] Available at http://www.barentsinfo.fi/docs/Gorbachev_speech.pdf [accessed: 10 June 2012].

Gottman, J. 1973. *The Significance of Territory.* Charlottesville: University of Virginia Press.

Government of Canada. 2010. House of Commons Standing Committee on National Defence, 40th Parliament, 3rd Session. Canada's Arctic sovereignty. [Online, June] Available at: http://www.parl.gc.ca/HousePublications/Publication.aspx?DocId=4486644&Language=&Mode=1&Parl=40&Ses=3&File=21 [accessed: 10 June 2012].

Harley, J.B. and Woodward, D. 1987. *The History of Cartography, Vol. 1: Cartography in Prehistoric, Ancient, and Medieval Europe and the Mediterranean.* Chicago: University of Chicago Press.

Hartshorne, R. 1953. Where in the world are we? Geographic understanding for political survival and progress. *Journal of Geography*, 52, 382–93.

Hau'ofa, E. 1993. Our sea of islands. In *A New Oceania: Rediscovering Our Sea of Islands*, edited by E. Waddell, V. Naidu, and E. Hau'ofa. Suva, Fiji: University of the South Pacific, 2–16.

Kitchin, R. and Dodge, M. 2007. Rethinking maps. *Progress in Human Geography*, 31, 331–44.

Latour, B. 1987. *Science in Action: How to Follow Scientists and Engineers through Society*. Cambridge, MA: Harvard University Press.

Lewis, M.W. and Wigen, K. 1997. *The Myth of Continents: A Critique of Metageography*. Berkeley: University of California Press.

Lewis, M.W. and Wigen, K. 1999. A maritime response to the crisis in area studies. *Geographical Review*, 89(2), 161–8.

Marston, S.A. 2000. The social construction of scale. *Progress in Human Geography*, 24(2), 219–42.

Mignolo, W. 2003. *The Darker Side of the Renaissance: Literacy, Territoriality, and Colonization, 2nd edition*. Ann Arbor: University of Michigan Press.

Mignolo, W. 2005. *The Idea of Latin America*. Oxford: Blackwell.

Ober, F. 1904. *Our West Indian Neighbors: The Islands of the Caribbean Sea, 'America's Mediterranean', their Picturesque Practices, Fascinating History and Attractions for the Traveler, Nature Lover, Settler, and Pleasure Seeker*. New York: James Pott & Co.

Semple, E.C. 1911. *Influences of Geographic Environment*. New York: Henry Holt.

Silva Greusz, K. 2006. The Gulf of Mexico system and the 'Latiness' of New Orleans. *American Literary History*, 18, 468–95.

Smith, N. 1987. Dangers of the empirical turn: Some comments on the CURS initiative. *Antipode*, 19(1), 59–68.

Smith, N. and Katz, C. 1993. Grounding metaphor: Towards a spatialized politics. In *Place and the Politics of Identity*, edited by M. Keith and S. Pile. London: Routledge, 67–84.

Stefansson, V. 1921. *The Friendly Arctic: The Story of Five Years in Polar Regions*. New York: Macmillan.

Stefansson, V. 1922a. *The Northward Course of Empire*. London: George G. Harrap.

Stefansson, V. 1922b. The Arctic as an air route to the future. *National Geographic Magazine*, August, 205–18.

Steinberg, P.E. 2001. *The Social Construction of the Ocean*. Cambridge, UK: Cambridge University Press.

Steinberg, P.E. 2007. Bridging the Florida Keys, in *Bridging Islands: The Impact of Fixed Links*, edited by G. Baldacchino. Charlottetown, PEI: Acorn Press, 123–38.

Steinberg, P.E. 2011. The Deepwater Horizon, the *Mavi Marmara*, and the Dynamic Zonation of Ocean-Space. *Geographical Journal*, 177(1), 12–16.

Steinberg, P.E. 2013. Of other seas: metaphors and materialities in maritime regions. *Atlantic Studies*, 10(2), 156–69.

Steinberg, P.E. and Chapman, T.E. 2009. Key West's Conch Republic: Building sovereignties of connection. *Political Geography*, 28(5), 283–95.

Wilkinson, K. 2012. Holland's 'Gulf of America' proposal: leave the jokes at home, say local lawmakers. *Gulflive.com* [Online, 11 February, updated 20 March] Available at: http://blog.gulflive.com/mississippi-press-news/2012/02/hollands_gulf_of_america_propo.html [accessed: 10 June 2012].

Zellen, B. 2008. We should warm to the idea of melting poles. *The [Toronto] Globe and Mail* [Online, 28 April] Available at: http://www.theglobeandmail.com/commentary/we-should-warm-to-the-idea-of-melting-poles/article1054687/ [accessed: 10 June 2012].

Chapter 3

'Plenty of Weeds & Penguins': Charting Oceanic Knowledge

Anne-Flore Laloë

Introduction

The exploration and mapping of the ocean's surface and the deep sea have always relied heavily on a number of technologies. Historically, these include compasses, sextants and octants, sounding devices and bathyscaphes. In recent decades, these have been replaced by their modern counterparts: radars, sonars and Niskin bottles, among others. Together, such technologies enable humans to overcome their inability to walk on water or remain under it for long periods, and thus facilitate the study of the ocean's surface and submarine life. On the whole, these devices have been examined by maritime historians and historians of technology as part of the chronology of maritime exploration, and within the enterprise of refining data collection and the production of knowledge about the ocean. In such studies, however, the ship, which is both a vehicle and a platform from which to study and know the ocean and its properties, has not typically been thought of in terms of its immediate contributions to scientific knowledge production. This is only now being addressed. As such, the ship's role as a scientific instrument, a laboratory and a space of science is being established both by historians and geographers (see Goodwin 1995, Rozwadowski 1996, Sorrenson 1996, Dunn and Leggett 2012). This work is helpful in understanding ships' importance in the exploration of and knowledge production about the ocean. However, in spite of recent attention towards the ship itself as a way of knowing the ocean, the relationship between ships and the traces they leave on charts remains unconsidered. The study of charts and tracks reveals complex relationships between ships, the ocean and geographical knowledge, and these relationships challenge certain notions of the ship as a scientific instrument and a space of science. This chapter will examine ways in which ships' tracks inform, or indeed fail to inform, knowledge about the geographical ocean. It will focus specifically on the discrepancy that is caused by using a fixed-grid graticule to map a moveable feature such as the ocean's surface, and the inherent difficulty of studying and mapping a constantly mobile space (the ocean) from a space that is physically distanced from it (the ship). I will first examine the spatial perspective offered by ships and how they affect knowledge production about the ocean. I will then examine a specific example, the HMS *Julia*, as it searched for the island of

St Matthew in the nineteenth century. I next consider what ships' tracks, and the annotations of these tracks, contribute to the charting of the ocean. To conclude, the chapter will consider the space of the ship itself in the process of mapping, and assess how it plays a role in geographical knowledge production.

Shipped Perspectives

The ship's importance in the exploration of the ocean is undeniable; it has repeatedly provided a platform for the ocean's study by navies, scientists and travellers alike. The fact that our perspectives of the oceans are 'shipped' ones then, is unsurprising.[1] Indeed, cartographers and explorers mapped the ocean from the perspective of ships and as such, geographical knowledge about the ocean is closely connected to the lines, or tracks, written using navigational tools and techniques, which represented the paths of ships across the ocean.

By the sixteenth century, a number of navigational methods were widely used in the West.[2] One was determining latitude – which was easily measurable, deduced by measuring the angle between the horizon and a fixed point in the sky.[3] Ships thus sought to remain on a specific line of latitude, or between two known lines. This method eliminated one unknown co-ordinate, though remained firmly set in a land-based perspective, as ships advanced on a line that, by its fixed and geometric nature, did not account for the ocean's moveable characteristics.

Another method of navigation was to stay near coasts and navigate following known features on land. This technique, though very efficient in well-mapped coastal areas, was limiting for exploration, though it did not preclude it. For example, through close observation and making drawings, sailors were able to extrapolate knowledge about the ocean, or indeed shape the geographical physicality and imagination of the ocean and specific ports of call. This was demonstrated in relation to the waters of Rio de Janeiro and the drawings of John Septimus Roe, which provide a visual history of the harbour (Driver and Martins 2002). This aspect of hydrographic mapping twinned with an analysis of the visual culture of imagination in the nineteenth century provides a useful insight into the challenges of navigating near coasts. Furthermore, it gives a place to the material ocean which creates distance from the coast and specific, observable qualities

---

1   This, however, is not unproblematic because a ship remains divorced from the ocean's physical characteristics: the perspectives it offers cannot fully speak to the ocean's space, since the ship is definitionally separated from the ocean itself.

2   This chapter does not discuss non-European modes of navigation, cartography or representation of the ocean. These have very different approaches to mapping the ocean, some of which reflect the ocean's physical properties more closely. Such a discussion, however, does not fall within the remit of this chapter.

3   However it should be noted that latitude only provided a partial position, which was not necessarily very useful for long-distance navigation or exploration.

that affect the ships' movements on the water or shape the viewer's perspective. Nonetheless, it remains that coastal navigation is limiting for journeys that seek to sail beyond the sight of land.

Moving away from the shoreline, the technique of dead reckoning is theoretically simple but relies on precise knowledge of oceanic phenomena such as tides and currents. These are now relatively well understood, are usually cyclical and predictable, and have been mapped precisely. This means that these phenomena can be used to accurately chart a vessel's route across oceans with utmost precision. Over short distances, positioning by dead reckoning is typically done hourly, though on longer journeys, points are typically marked daily as seen in Figure 3.1. The lines between two points made 24 hours apart are not geographically linked to the physical route of the ship, which more than likely tacked a number of times during the day. Indeed, the traces left by dead reckoning on charts at this scale, today or in the eighteenth century, are not representative of a ship's geographical position but are a smoothed representation of a much more saccadic movement. What they are most representative of is the problem of charting a ship's course, highlighting the difference between a course and a drawn track, and therefore bringing to the fore the problem of using them to deduce geographical knowledge.

These lines represent first and foremost a temporal relationship with the ocean, not a spatial one. In other words, the positions they mark must primarily be understood in relation to other points marked on the chart and as representative of a chronological flow of events, not as geographical place markers. What these points and lines show is that, at a certain point, after having followed a certain bearing, a ship is estimated to be at a particular relative point. This is because each point is drawn in relation to all the previous ones, so these lines do not represent a strict geographical co-ordinate that defines a unique locale, as is typically understood by location points, but a temporal succession of relative positions.

It is here that it starts to become clear that thinking geographically about the traces left on charts by ships is not straightforward. It is very difficult to map the ocean's surface, and it is impossible to do so in a manner that is most appropriate for mapping terrestrial surfaces. Mapping a space that is constantly in motion cannot be done in the same way that immobile land is, and spatio-temporal hints, such as tracks and their annotations, become important geographical clues. Thus, gleaning information from a 'shipped perspective', especially given the ubiquity of ships' tracks on maps and charts, presents issues regarding the production of knowledge about the ocean as a space.

Indeed, like all other physical geographical features, the ocean is defined by specific material and spatial characteristics. These include, among other qualities, the regular cycle of tides, the global network of transoceanic currents and water temperatures linked to water depth. Because these spatial and material features are visible and tangible within human timescales, that is hours and days rather than millennia, it would be limiting to study the ocean's space without bringing to the fore these qualities. In addition, since the ocean's physical characteristics are key in shaping, usually limiting, human interactions with the ocean at all levels, they cannot be excluded from geographical discourses. This is something which I have argued elsewhere, stating that the geography of the ocean must put at its core these essential characteristics and more fully acknowledge them, rather than apologize for them (Laloë 2009). This stance is also adopted by Steinberg and Bischof (2012) who posit that we should put 'marine metaphors to work' and align maritime geographical scholarship with the challenges faced by oceanographers in their scientific endeavours. In this way, the ocean's characteristics become integral to thinking geographically about the ocean and these in turn become crucial in considering how the ocean is charted, usually in manners that fail to encapsulate their moveable characteristics. Whereas this is, to a certain extent, being remedied by recent developments in data visualization and cartographic representation which can be dynamic rather than fixed, historical charts of the ocean represent a two-dimensional space that is difficult to reconcile with the physical reality of the material ocean.

Moreover, if points and tracks cannot be interpreted directly within the context of the geographical graticule but need to be analyzed in their locality and within a relational timeframe, the paradigms of place, where place is defined as a unique, physical locale (a definition at the core of a traditional understanding of geographical knowledge production), find themselves shifted. Place and geographical knowledge about the ocean become instead centred upon a ship's perspective within a moveable and time-dependent frame. This becomes more apparent when investigating specific charts and thinking about the place of the ship in affecting the way the ocean is known. Introducing these maps here will allow for a more specific discussion on ships and geographical space.

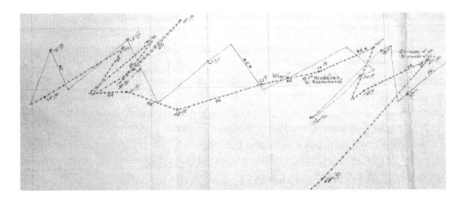

Fig. 3.1   **UKHO Survey C. 138:** *The Track of His Majesty's Sloop Ship Julia, Jenkin Jones Esqr. Commander, in search of the Island of St Matthew,* **1817. Reproduced in part from the collections of the United Kingdom Hydrographic Office.**

Tracking *Julia*

In 1817, HMS *Julia* went on a month-long expedition zigzagging along the second line of latitude south of the Equator. The journey is charted on *The Track of His Majesty's Sloop Ship* Julia of 1817.[4] The journey charted reads as follows: On 16 September 1817, *Julia* departed Saint Helena and made her way to, approximately, 2 degrees 30 minutes of latitude south and 1 degree of longitude west, where she arrived on 21 September. At this point, she steered westward, meandering to the meridian of 14 degrees longitude west, before turning back on 4 October, apparently seeing a rock on 15 October, and heading southwards towards Ascension Island. *Julia* arrived at Ascension Island on 21 October 1817 after 36 days at sea.

The purpose of this journey was the geographical exploration of a portion of the equatorial Atlantic Ocean, in the area now known as the Gulf of Guinea. *Julia* was searching for an island called St Matthew which had been on nautical charts since García Jofré Loyasa had drawn it in 1525. Loyasa was on his way to the Moluku Islands in Indonesia when, according to where he thought he was, he saw something on the horizon. A few days previously, he had crossed a Portuguese ship whose crew had given him directions to what they said was a nearby, little-known island. Thus, in October 1525, Loyasa then set foot on 'a high land, four leagues around, covered with palms and oranges. There were no inhabitants, only human bones upon the ground, remains of houses, and a wooden cross' (Strommel 1984: 121). Since there was nothing in that position on the chart he was using, Loyasa dutifully added the island on at his assumed position approximately 2 degrees of latitude south and 8 degrees of longitude west, and named it the Island of St Matthew. Despite the lack of further evidence of the existence of St Matthew, the island continued to appear on maps and surveys until the nineteenth century. Numerous expeditions, including the ones undertaken by *Julia*, HMS *Inconstant* or the French vessel *Cigogne*, set out to find the island with the purpose of exploring it further. Since it is now known that St Matthew does not exist, it is not surprising that the island remained unfound. As this became apparent, the island disappeared from charts. What remained, however, were tracks on maps. These illustrate the difficulty of charting the ocean's surface and are important proxies of geographical knowledge.

The tracks that were traced are now seen as revealing a complex relationship between the ship, the track and the oceanic space. The path that *Julia* actually travelled is twice removed from what is drawn on chart. This distancing is due to the challenges of dealing geographically with the material ocean. What *Julia* reveals is a useful example of both the challenges and discrepancies that are at the core of mapping the ocean from a ship's perspective. Thus, if the chart is to be used to extrapolate knowledge about the ocean's geography, we need to address the question of what, if any, geographical knowledge can be deduced from such charts and their tracks. One way of addressing this question is through the annotations adjoining features or position marks. These annotations add a new layer to the relationship between the ship, the perspective it offers and the ocean.

---

4 UKHO Survey C. 138: *The Track of His Majesty's Sloop Ship* Julia, *Jenkin Jones Esqr. Commander, in search of the Island of St Matthew*, 1817.

Spaces of Seaweed and Penguins

One of the particularities of *Julia*'s chart is that it revolves around the search for a non-existent island. That St Matthew did not exist was not known at the time of the journey, but its assured presence on the chart is telling. St Matthew appears on this chart in only one position, approximately at the place Loyasa had located it. *Julia's* search is focused on exploring the island's latitude line (explained by the fact that, as described previously, it was relatively straightforward to measure latitude); therefore, if one thing could be trusted from Layosa's chart it was the given latitude of St Matthew. Longitude was the more problematic co-ordinate, which explains why *Julia* sought to cover as much of it as feasible.

The case of *Julia* and St Matthew, however, is not atypical, and the history of mapping the ocean is peppered with islands that were wrongly located and repeatedly moved before being firmly positioned in a fixed geographical place or, as in the case of St Matthew, deleted when their non-existence was ascertained. To a certain extent, the veracity of the islands themselves becomes irrelevant: what is especially curious is the visual methods used to represent these ambiguity of place in these instances.

In the area of the Falkland Islands, Isle Grande is mapped on the 1807 *Chart of South America and the Southern Ocean* and the 1808 *Chart of the Ethiopic or Southern Ocean, and part of the Pacific Ocean* (Figure 3.2).[5] On the former of these two maps, the Isle Grande is positioned in three different places, all on the same line of latitude but a few degrees of longitude apart. Snaking around the island's alleged positions are the tracks of two ships: that of Monsieur de la Pérouse in 1785 and that of Captain Vancouver in 1795. Both had captained ships searching for Isle Grande. The captions next to each location read thus: 'I. Grande according to M^r. Dalrymple', 'Probable situation of the I. Grande; but very uncertain' and 'I. Grande. A good harbour. Discovered by La Roche in the year 1675 and as laid down by Capt. Cook's, but erroneously, as it appears by the track.' The track referred to here is a section of La Pérouse's search for the island in 1785. On the 1808 map, the cartographer resorted to the same technique of drawing the island three times, annotating each time to indicate the uncertainty of his claims. The Isle Grande locations are marked, again, to reflect the origin of the source thus: 'I. Grande according to Dalrymple', 'I. Grande. A good harbour. Discovered by la Roche in 1675. Situation very uncertain' and 'I. Grande. According to Capt. Cooks [sic] chart.'

---

5   UKHO Survey B. 424 5. G: *Laurie and Whittle's Chart of South America and the Southern Ocean; Including the Western Coast of Africa, from Cape Verd* [sic] *to the Cape of Good Hope,* 1805; UKHO OCB. 357 A. 1: *A Chart of the Ethiopic or Southern Ocean, and part of the Pacific Ocean; from the Parallel of 3 degrees North to 56° 20' South Latitude and from 20° East to 90' West Longitude drawn from the latest observations of the Spanish, Portuguese and Dutch Astronomers, shewing the track of the Warley, East Indiaman, outward and homeward in the years 1805 & 6,* 1808.

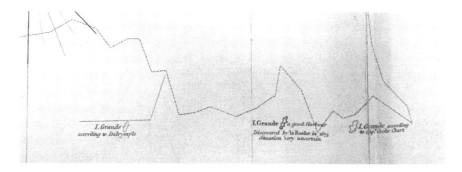

Fig. 3.2 **UKHO OCB. 357 A. 1:** *A Chart of the Ethiopic or Southern Ocean, and part of the Pacific Ocean; from the Parallel of 3 degrees North to 56° 20' South Latitude and from 20° East to 90' West Longitude drawn from the latest observations of the Spanish, Portuguese and Dutch Astronomers, shewing the track of the Warley, East Indiaman, outward and homeward in the years 1805 & 6,* **1808. Reproduced in part from the collections of the United Kingdom Hydrographic Office.**

These annotated tracks bear witness to the complexity of affixing places to oceanic spaces. It also highlights how cartographers and explorers were influenced by the Enlightenment and particular ways of knowing the world. Isle Grande's hypothetical locations and the tracks marked on the charts, indicating a location, produce a place on the ocean – thereby furthering knowledge production. Despite the uncertain nature of the knowledge being produced, the uncertainty nonetheless reflects an *ideal* of knowing the ocean in strict, 'Enlightened' ways. In this sense, the tracks on the maps and the triple representation of the Isle Grande illustrate the principles of corrective science that was at the core of the Enlightenment's mission. The challenges posed by the material ocean put emphasis on this process, since a straightforwardly chartable surface would not require the multiple positioning of islands.

In order to fully engage with the ocean's physical characteristics, the question of what annotations mean in terms of geographical knowledge remains central. In other words, what does it mean that a ship has or has not seen an island one day, somewhere on the vast oceanic surface? Can reliable geographical knowledge, such as the existence of an island or a continent, be inferred? For instance, it was from the tracks of HMS *Resolution* (1772–1775) in the Southern Ocean that James Cook was able to conclude that, were a southern continent to exist, it would be located south of his vessel's track, probably on the South Pole itself. This kind of geographical deduction evidences a very specific process of knowledge production where the ship and its track are tightly intertwined with the nature of geographical knowledge at sea.

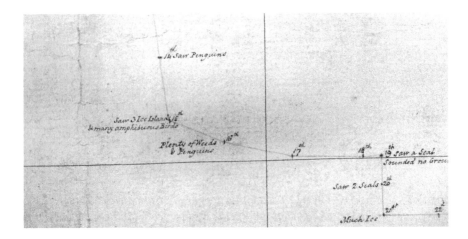

Fig. 3.3   UKHO Survey Z1 (shelf Hf): *Carta Esferica del Oceano Meridional desde el Equador haste 60 grados de latitud y desde el Cabo de Hornos hasta el Canal de Mozambique construida de orden del rey en la dirreccion de trabajos hidrographicos y presentada á S. M. por mano del excmo. señor Don Antonio Cornel, Descretario de Estado, y del Despacho Universal de Guerra, encargado del de Marina, y de la direccion general de la Armada, año de 1800.* **Reproduced in part from the collections of the United Kingdom Hydrographic Office.**

For example, looking at the extract from the *Carta Esferica del Oceano Meridional desde el Equador* (Figure 3.3), a mark on a chart, which notes that on the 16th of a date unspecific on this chart (though this information might be gathered form associated logbooks), a ship charting its route by dead reckoning in the southern Atlantic Ocean, saw 'plenty of Weeds & Penguins' can pose more questions than it answers.[6] The only facts that are known, from looking at this map, are that in a theoretically calculated, relative position, weeds and penguins were seen from a ship, sometime between two points in time over a 24 hour watch. It is apparent that this, on its own, is not particularly useful information in terms of geographical knowledge production, or even simply recording a specific ship's journey. While the inanity of what is recorded is perhaps comical in isolation, the fact that this note exists is a testament to the unique geography of the ocean and

---

6   UKHO Survey Z1 (shelf Hf): *Carta Esferica del Oceano Meridional desde el Equador haste 60 grados de latitud y desde el Cabo de Hornos hasta el Canal de Mozambique construida de orden del rey en la dirreccion de trabajos hidrographicos y presentada á S. M. por mano del excmo. señor Don Antonio Cornel, Descretario de Estado, y del Despacho Universal de Guerra, encargado del de Marina, y de la direccion general de la Armada, año de 1800.*

the historical geography of the ocean through charts. In this sense, considering the disconnection that exists between an annotation, the ship whence the observation noted was made, and the time and place that are recorded in the co-ordinates, each element appears to be inherently alien to the ocean. Yet, it encapsulates the challenges of charting the ocean and provides a helpful perspective regarding the geographies of the ship as a space of geographical knowledge production *at* sea, which forges knowledge *about* the sea.

Conclusions

The spatiality of ships is multifaceted. Foremost, ships are themselves physical spaces that are imagined, designed, built, lived in and worked on. Intrinsic to their layout and qualities, however, is the fact they are spaces that are designed to navigate on the ocean, a moveable, uncertain surface which is itself spatially complex. Furthermore, the ship, as a place, is easily thought of, or even defined, by the actions that happen on board, and these shape ships' and crews' geographical interactions with the ocean. This is something which geographers have begun to establish and which is at the core of current geographical studies of the sea. What is emerging is that if geographers struggle to speak about the geographies of the ship, it may be because there are as many geographies of ships as there are ships, and each of these create unique geographies of the ocean's space.[7] This important point allows us to think about the ship not as an abstract space, but one which offers useful perspectives on geographical knowledge production at sea.

Thinking of the ship as a space of science offers a useful way of addressing the relationship between knowledge production and ships and ultimately, our understanding of the oceans. Following the work of historical geographers and historians of science, the notion that science is situated knowledge enables us to think geographically about the spaces of science and consider scientific practices and places as integral to science itself (Haraway 1988).[8] Due to the spatial and geographical characteristics of both the ship and the ocean, the localities charted by *Julia*, for example, are geographically challenging and intrinsically removed from the traditional ideas of place as fixed to specific co-ordinates. Yet, at the same time, they are the only geographical markers available for thinking geographically. While, following Withers (2007: 97), *Julia*'s journey fits within the Enlightenment's travel mission of 'bringing the world to light less by imposing a single universal standard – than by calibrating others local standards with a view to ensuring, in time, a commensurability over space', it is important that we consider the spaces where this calibrating of science occurred, and the notion of the ship as a scientific instrument is particularly helpful here.

---

7   This statement on the ship geographies results from discussions held a round table event at Royal Holloway, University of London on 8 March 2011.

8   See also Secord 2004, Latour 1987, Latour and Woolgar 1986, Livingstone 2003, Livingstone and Withers 1999, Withers 1995, 2006 and 2007.

The idea that the ship is a scientific instrument was proposed by Sorrenson (1996) when discussing the ship's role in answering questions about the physical nature of the globe. Sorrenson argues that the scientific ship is at the core of the Enlightenment's geographical enterprise, enabling transport across the ocean's surface while at the same time facilitating the exploration of the Earth. This latter point brings to the fore the ships' individual tracks and how the mapped exploration of the Earth, in turn, shapes our knowledge of the Earth. The tracks, then, demonstrate how an actual, individual ship can impose its identity onto the ocean. That ships, unlike other modes of transport, leave marks on charts is distinctive, since it not only (attempts) to locate (landed) places, but also reveals other stories linked to those particular places, all the while, paradoxically, not revealing much about the ocean itself.

Indeed, ships' tracks are deeply reminiscent of what MacDonald (2006: 642) calls the '"placeless" mobility' of the ship. In the case of *Julia*, the tracks and ship confirm that there was no island where or around where the ship had theoretically been. In relation to the production of geographical knowledge, the ocean's physical characteristics and the challenges of identifying and recording location on the ocean's surface, highlight the contradictory nature of the ship as a 'scientific' instrument. Indeed, the ship's placelessness on the ocean surface (in other words, its theoretical location drawn à posteriori on a chart) is at odds with the geographical and spatial imperatives of thinking of the ship as a scientific instrument – a space of science. Ships' tracks bring to the fore the practical complexities of recording localities on the ocean's surface and the challenges of measuring position while at sea. The annotations that caption the tracks either provide provenance for geographical knowledge, especially if it is questioned, or reinforce the ship's importance by locating a specific sighting or perspective. This jars, however, with the specific role of the ship and its role as a scientific space. Certainly, this is in part due to the fact that the physical characteristics of the ocean do not sit easily within the land-based geographical graticule, which means that charts and traces on charts are intrinsically at odds with what is typically encountered at sea. However, using ships and charts to the study of the ocean, where the ocean, ships and tracks are physically bound to each other, creates a unique method of critically unpacking how our understandings of ocean space are formed, and, as shown in this chapter, further offers the potential to reform those understandings.

The place of the ship in geographical knowledge production has been crucial here. What the traces left by *Julia* exemplify is that it is only by examining the ships' tracks and the thinking about the ships that (theoretically) made them that it becomes possible to understand how tightly intertwined they are with the physical ocean. The tracks of *Julia* are representative of the geographical complexity of mapping the ocean. In fact, they can be said to encapsulate the very nature of making geography at sea, bringing to the fore the complex spatial and geographical relations between the ship as a vehicle, the ship as a space of science and the ocean itself, with its indeterminate, though evocative, weeds and penguins.

References

Bischof, B. and Steinberg, P. 2010. Fluid Dynamics and Dynamic Fluidity: Putting Marine Metaphors to Work. Paper to the Royal Geographical Society (with Institute of British Geographers) Annual Conference. London, 2 September 2010.

Driver, F. and Martins, L. 2002. John Septimus Roe and the art of navigation, c. 1815–1830. *History Workshop Journal*, 54, 144–57.

Dunn, R. and Leggett, D. eds. 2012. *Re-inventing the Ship: Science, Technology and the Maritime World, 1800–1914*. Aldershot: Ashgate.

Goodwin, C. 1995. Seeing in depth. *Social Studies of Science*, 25(2), 237–74.

Haraway, D. 1988. Situated knowledges: The science question in feminism as a site of discourse on the privilege of partial perspective. *Feminist Studies*, 14, 575–99.

Laloë, A.-F. 2009. Knowing the Ocean-Space: An Atlantic Case Study, Ph.D. thesis, University of Exeter.

Latour, B. 1987. *Science in Action: How to Follow Scientists and Engineers through Society.* Cambridge, MA: Harvard University Press

Latour, B. and Woolgar, S. 1986. *Laboratory Life: The Construction of Scientific Facts*. Princeton, NJ: Princeton University Press.

Livingstone, D. 2003. *Putting Science in its Place: Geographies of Scientific Knowledge.* London: University of Chicago Press.

Livingstone, D. and Withers, C. eds. 1999. *Geography and Enlightenment.* London: University of Chicago Press.

MacDonald, F. 2006. The last outpost of Empire: Rockall and the Cold War, *Journal of Historical Geography*, 32, 627–47.

Rozwadowski, H. 1996. Small world: Forging a scientific maritime culture for Oceanography. *Isis*, 87(3), 409–29.

Secord, J. (2004) Knowledge in transit. *Isis*, 95, 654–72.

Sorrenson, R. 1996. The ship as a scientific instrument in the eighteenth century. *Osiris* 2nd Series, 11(4), 221–36.

Strommel, H. 1984. *Lost Islands: The Story of Islands that have Vanished from Nautical Charts.* Vancouver, BC: University of British Columbia Press.

Withers, C. 1995. Geography, natural history and the eighteenth-century enlightenment: Putting the world in place, *History Workshop Journal*, 39, 136–63.

Withers, C. 2006. Eighteenth-century geography: Texts, practices, sites. *Progress in Human Geography*, 30(6), 711–29.

Withers, C. 2007. *Placing the Enlightenment: Thinking Geographically about the Age of Reason.* London: Chicago University Press.

Chapter 4

Geographies of Coral Reef Conservation: Global Trends and Environmental Constructions

Bärbel G. Bischof

Introduction

> The symptoms of environmental deterioration are in the domain of the natural
> sciences, but the causes lie in the realm of the social sciences and humanities.
> (Orr 1992: 146)

In the eyes of most marine scientists, oceans are undergoing a crisis. Current
narratives of environmental distress are concentrated mostly on large-scale
systemic changes, such as a warming and acidifying ocean, expanding eutrophic
coastal zones, prolonged periods of anomalous conditions, advancing sea-level
rise, pollution, and rapid attrition of global fisheries stocks (Burke et al. 2011,
Baker et al. 2008, Hughes et al. 2003, Kleypas and Yates 2009, Knowlton 2001,
2008, Roberts et al. 2002). Although these problems manifest in various ways
in ocean environments, the ecological complexities and geographies of coral
reefs makes these tropical systems particularly vulnerable to the repercussions
these environmental transformations could have. Because of their global
expanse and distribution, and the uniform way that reefs display environmental
stressors, these systems are collectively considered a comprehensive indicator of
global ocean health, and the prognosis for their future is unanimously seen as
bleak (Anthony et al. 2008; Burke et al. 2011, Dimitrov 2002, Ginsburg 1994,
Kleypas and Eakin 2007, Kleypas and Yates 2009, Knowlton 2008, Reakla-Kunda
1997). With institutional systems firmly in place to advocate reef protection and
conservation, which includes user-groups, grass roots organizations, non- and
inter-governmental organizations (NGOs and IGOs) and governments claiming
a common goal of sustainability, a worldwide epistemic community of reef
researchers, and increasing coverage in the general press,[1] progress has not been

---

1 For example, International Coral Reef Initiative *Australia announces world's
largest marine reserve network* [Online] Available at: http://www.icriforum.org/
news/2012/06/australia-announces-worlds-largest-marine-reserve-network and NOAA

made in improving the global status of reef systems. This failure is attributed to the lack of clarity in scientific knowledge (Dimitrov 2002) but as I will argue, is also shaped by the beliefs and attitudes within the epistemic community of reef-researchers.

Despite the significant volume of scientific work in reef ecosystems that has accumulated over several decades, and the persistent narratives of ecosystem doom, scientists and reef professionals who work in these systems display an inordinate level of disagreement about how to slow (or halt) environmental decline. Competing statements regarding the most important reef-stressors, the most significant socio-cultural obstacles of conservation, the extent of the conservation problem, monitoring methods, the best potential mitigation strategies, and the significance of research efforts are among the most contested elements, all of which differ drastically by geography, disciplinary expertise, social networks, and years of activity in the field. These issues vary in importance depending on region and place, and also fuel dissonance among scientists about which features and dynamics are most urgent to address to mitigate trends of decline (Bischof 2010a, Delaney and Hastie 2007, Dimitrov 2002, King 2005, Kleypas and Eakin 2007, Mora et al. 2011).

Although all reef scientists and managers explicitly state a common goal of furthering conservation, the enormous set of variables and viewpoints are clearly not conducive to achieve the goal of finding policy solutions or management regimes that are widely supported (Dimitrov 2000, West and Brockington 2006). The most imposing problem that prohibits a comprehensive, co-ordinated reef regime is that disagreement and preferencing of competing truth-claims within a social framework does *not* occur within an objective evaluation of a big picture that considers all options, but rather rests mainly on the social contexts, associated networks, technologies employed, and the spaces in which the claims originate (Bischof 2010a, Latour 1987, Dimitrov 2002, Powell 2007). The narratives tied to the processes of negotiating elements of fact and generating scientific discourses are thus direct reflections of rudimentary subjectivities within the epistemic community, specifically deep-seated attitudes and beliefs, ideal imaginations of environment, which are inevitably biased by spaces of science (Livingstone 2003, Powell 2007, Stoddard 1986). In this chapter I explore an alternative approach to gain new insight into how to circumvent uncertainty and subjectivity in ocean environments, in this case the crisis of coral reefs.

I start this chapter by providing a background to coral reef environments, exploring how they are a sign of global ocean health – the proverbial 'canary in the coal mine' – investigating the ways in which such ecosystems have been threatened, driving a growing conservation rhetoric to protect them. I next explore Marine Protected Areas (MPAs) as a key conservation strategy, tracing the challenges in attempting to manage these vulnerable areas due to competing science claims alongside socio-cultural and contextual differences in the geographies of

---

Coral Reef Conservation Program [Online] Available at: http://coralreef.noaa.gov [accessed 14 June 2012].

reef systems. I then suggest an alternative way in which to assess the conflictual narratives underscoring conservation of reef systems, using Q methodology. Following this, I present results, finding trends of consensus and difference in how such ocean environments are imagined and understood, before noting how this links to the real-world management of the coral reef ecosystem.

By identifying geographies of disagreement in terms of dominant narratives of environmental problems and solutions for reef conservation, the geographic patterns of crisis and mitigation discourse become evident, as do the core sources of tension in the coral reef conservation crisis. Outlining these discursive paths demonstrates how our knowledge and understanding of the ocean directly influences how we actually manage ocean space. Ocean narratives and the paradigms that emerge from them are not intangible; they have very real consequences in shaping how we engage with and act upon its systems.

Background: Geography of Coral Reefs

Hermatypic corals build reef systems that are complex, near-coastal ecosystems concentrated in a belt straddling the equator, generally confined to clear, warm, shallow oceans between 25°N and 25°S (Figure 4.1). The reef structure is made of limestone that the legions of (mostly) colonial corals precipitate out of the carbonate-saturated waters of the tropics, using sunlight for the necessary energy with which to construct their habitats. The complex ecological web that makes up a reef complex contains many trophic levels and boasts the highest biodiversity— by phyla—of any other ecosystem on Earth (Carr et al. 2003, Riegel et al. 2009). As limestone producers, corals are sensitive to pH and temperatures. The energy they need to make their structures is provided by photosynthesizing symbionts living in their tissue, single-celled plants called zooxanthellae, which give coral polyps their vibrant colors, while also restricting growth to the photic zone.

Reefs provide valuable ecosystem services that reach far beyond their location. Although covering less than 1% of the surface of Earth, an estimated one-third of all marine creatures known, call reefs home at some point in their life-cycles, including 25% of all seafood species consumed (Caddy and Garibaldi 2000, Dimitrov 2002). Because of their particular growth preferences in shallow zones, and their coincidental location in hurricane-prone regions, the coral structures have been documented to act as natural breakwaters, reducing destructive storm-surges and coastal erosion to otherwise exposed shorelines and reducing costs of post-storm recovery, as well as saving countless lives and livelihoods (Burke et al. 2011, Guzman et al. 2003).

An estimated 61% of coral reefs are experiencing local stresses that put them at serious risk of eradication by 2030 (See Figures 4.1 and 4.2) (Wilkinson 2008). Thus although reefs are arguably a particular and unique system in terms of geographic extent and environmental conditions, they are nevertheless a key ecosystem for the planet. In fact, these systems are understood to be among

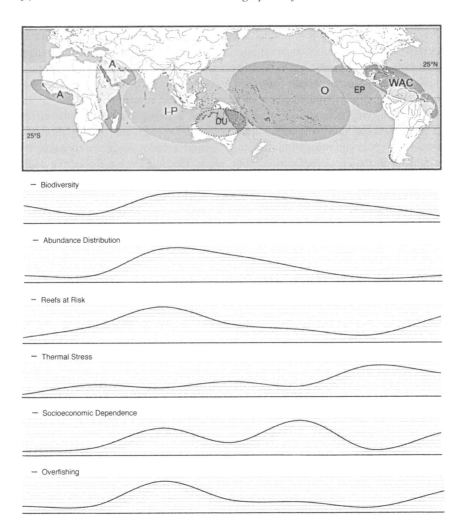

Fig. 4.1   Geographic distribution of reefs by region.

Note: A=African coast and the Middle East (very few reefs are found in the eastern Atlantic); I-P=Indo-Pacific; DU=Australia; O=Oceania, which includes the South Pacific and Hawaii; EP=Eastern Pacific; WAC=Western Atlantic and Caribbean. The line graphs below show the relative intensity/urgency of the particular variables or characteristics by region as described by major peer reviewed scientific articles. E.g., the highest biodiversity is in the Indo-Pacific region, particularly the Indonesian/Malaysian archipelagos, and Papua New Guinea, while the least biodiversity is in the Western Atlantic/Caribbean region.

Source: Bischof 2010a.

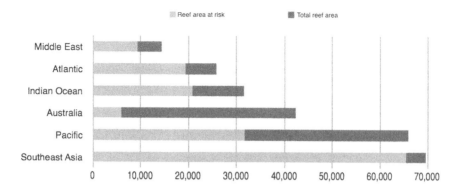

Fig. 4.2   Reef area under threat by region.

Note: Bars represent the total area of reefs (in square kilometres) for each region. Light shaded area represents the total reef area undergoing medium or higher threat levels from the variety of stress. These include coastal development, watershed-based pollution, marine-based pollution, overfishing and habitat destruction, thermal stress, and ocean acidification.

Regional scope: Atlantic=Caribbean, North America and Eastern South America; Australia= separated from other reef regions because of its development status and available resources for protection; Indian Ocean=includes all reefs bordering the Indian Ocean basin except those belonging to North Africa and the Middle East; Middle East=Red Sea and Persian Gulf reefs, and the Arabian Peninsula; Pacific=all reefs in the Philippine Sea and eastward to the Galapagos and western shores of the Americas. Southeast Asia=reefs found west of the Philippine Sea, east of the Indian Ocean and north of Australia. Distribution, in percent, of the estimated 275 million people who live within 30 kilometres of a coral reef.

Sources: Burke et al. 2011 and Bischof 2010a.

the most obvious 'canaries in the coal mine' for oceans, and some would argue global environmental health in general, because of the visible way in which they show signs of stress (Birkeland 1997, Burke et al. 2011). When temperatures or water-quality conditions become intolerable, corals will expel their zooxanthellae symbionts. Without these pigmented plants in their clear, mucus-like tissue covering the colonies, their limestone skeleton beneath becomes visible and they appear a glaring white, a phenomenon known as bleaching. Because of the role of these ecosystems in supporting valuable species of fish and invertebrates, their efficient ocean geochemical recycling abilities, and prolific primary productivity when healthy, reef systems are also called 'rainforests of the sea'. Although only about 93,000 reef species have been described and catalogued, estimates for the actual number of possible species that reside on reefs ranges from 950,000 to 9 million. Concerns regarding impeding potential biodiversity are also growing over the unknown consequences of increasing regional mass extinctions of associated, linked species and ecosystems (Knowlton 2008, Reakla-Kudla 1997).

Moreover, reefs are found in jurisdictions of an enormous variety of polities in various stages of development and are thereby subjected to a diverse set of cultural norms and geographies. A little over 100 countries have coral reefs, although more than half of them are located within the jurisdiction of only six states: Australia and Indonesia (which have 35% of global reefs within their waters); Fiji, the Maldives, Papua New Guinea and the Philippines (sharing 15%); and the rest sprinkled throughout the remainder of reef-regions (Burke et al. 2011, Dimitrov 2002, Spalding et al. 2001). Currently, the total area of formally protected reefs in the form of marine protected areas (MPAs) stands at about 27% of reefs worldwide, although it is estimated that less than 6% of MPAs are actually effective. This discrepancy is attributed to problems relating to policy enforcement, poaching, habitat destruction from externalities of development (such as land-based sources of pollution and sedimentation), resource exploitation (e.g. large-scale, off-site commercial fisheries), and other socially-determined stressors that connect to deeper issues like poverty and growing disparities of opportunity and wealth (Burke et al. 2011, Hughes et al. 2002, Mora et al. 2006).

Yet despite concerted efforts for many decades to protect reef systems[2] there is escalating urgency in conservation rhetoric, fueled by evidence of rapid rates of ecosystem decline across the globe. The first marine park, John Pennekamp State Park in Key Largo, Florida, was established in 1963 and the Great Barrier Reef off of Eastern Australia in the mid-1970s. It is generally appreciated that these systems entered the public eye through Jacques Cousteau's pioneering undersea videography from his first *Calypso* expedition to the Red Sea in the 1950s and 60s. Broadcast in the late 1960s on television, images from Red Sea reefs in Cousteau's *The Silent World* presented these systems as precious, unique, vibrant, and crucial to healthy seas and human existence and framed them as a co-dependent system, intricately linked to the health of an ocean world that positioned society as their stewards. Spectacular scenes of never-before filmed reef creatures were burned into the imaginations of an entire generation of young people, motivating them into what was then seen (and popularized) as a glamorous career in ocean science (Bischof 2010a, Ginsburg 2012, Kroll 2008). Throughout the 1970s, safer travel, better connections to coasts, exponential growth in accessibility and popularity of scuba diving, and well-funded marine science efforts would make reef conservation a global narrative by the early 1980s, generating a surge for the already burgeoning field of marine science (Ginsburg 1994, 2012).

Despite the turn of understanding ancient reefs as potential resources, a conservation emphasis of living reef systems dominated in the public sphere in the 80s and echoed throughout the 1990s in the form of dozens of new NGOs and IGOs that sprang up to establish and manage the flurry of new MPAs. About 40 years and 2,800 MPAs later, efforts to halt increasing rates of decline in these key

---

2   This began with the general trend of late 1960s environmentalism movements – see Birkeland 1997, Bowler 1992, Burke et al. 2011, Knowlton 2008, Reakla-Kunda 1997, Souter and Linden 2000.

ecosystems has still not been successful, in fact are sometimes framed as wasteful or pointless considering the significant scientific resources and devoted epistemic network that has been researching these systems over the past decades with no real progress (Burke et al. 2011, Dimitrov 2002, Kenchington 1998, Mora et al. 2006, Veron et al. 2001).

Marine Protected Areas (MPAs)

Devising any effective management system involves co-ordinating a complex set of variables (Pelletier et al. 2005); worldwide MPAs have made no significant headway in conserving reef ecosystems (Hargreave-Allen et al. 2011). Although reef scientists and managers can provide scores of empirical reasons for this overall failure, it is apparent, through a geographer's lens, that the human and physical geographies of reef areas vary far too greatly between and within reef systems to define a clear set of core characteristics that can be universally applied as a way towards mitigation of environmental destruction. Reef researchers and managers are well-aware that protected areas must be customized to suit proximal geographies if they are to have any chance of success, but in their implementation it has become evident that the particular policy-measures that work cannot always be correctly anticipated or forecast because of the intricate human-nature interactions that consistently influence the fate of these systems on local scales.

However the larger problem in MPA efficacy rests on the sub-text that these systems are designed around the basic premise of outsider territorial control and resource appropriation as the standard conservation solution (Burke et al. 2011, Hargreave-Allen et al. 2011). The creation of what essentially amounts to a semi-privatized, open-access system, as most MPAs are, has sparked user-conflict through the elimination of tenure systems, especially in poor, developing countries and small island-nations where long standing community taboos and local norms of resource stewardship are often the only functioning mechanisms of sustainable exploitation. Although these tenure systems are mostly within developing and impoverished countries, the formation of MPAs has resulted in increasing rates of reef decline because of the elimination of informal systems of community oversight and policies that are not compatible with local traditions (Burke et al. 2011, Jentoft et al. 2007, Hargreave-Allen et al. 2011).

A geographic inventory of the MPAs worldwide has shown that many of these sites are generally biased towards coverage on reef areas that are distant from high-use areas and the associated localized ecosystem pressures, and so play a minor role in halting declines or changing behaviors that may promote sustainable use because these locations were not particularly threatened to begin with (Burke et al. 2011). Genetic evidence countering the idea of inter-basin reef connectivity via ocean currents and revealing that most reefs are self-recruiting (i.e. larvae and fry usually settle on their native reef) has brought even more urgency to questions of scales of protection (Cowen et al. 2006) as well as the consequences of the activity spaces of reef fishers, and has brought increasing popularity to the notions of Large Marine Ecosystems (LMEs) as a better choice over localized MPAs (Sherman 1995); however this LME-scheme has also not proven successful because of its large scale, which concurrently invokes intractable issues of socio-political geographies and environmental justice given the breadth of political systems they encompass (Juda and Hennessey 2001, Longhurst 2003).

To their credit, however, some MPAs do seem to be working, however these are geographically specific and their policies cannot be exported to the myriad of troubled sites. The most effective MPAs are located along Australia's Great Barrier Reef, in atoll regions in Oceania, and the uninhabited north-western Hawaiian Islands. These MPAs have had some local successes in slowing coral loss, reducing destructive use practices, and increasing tourism and local employment, despite most being generally underfunded and facing large-scale threats that remain outside the control of park managers and local users (Burke et al. 2011, Hargreave-Allen et al. 2011, Mora et al. 2006). The common narrative to explain the success of these particular MPAs maintains that the health of any reef area is directly proportional to nearby population densities and resources for enforcement measures. The apparent correlation between reefs under threat and coastal populations, however, is accurate only for the extremes, and is non-existent for the Middle East, the Indian Ocean, and the Atlantic reef regions (Figure 4.3). Nevertheless, the promise shown by successful efforts has galvanized a significant number of reef scientists who staunchly insist that current ideas of MPAs as 'use-zones' do not make sense without also severely restricting large areas of ocean from human use, and will be the only way reefs will survive global change (Jackson 1997, Jackson and Sala 2001, McLanahan 2011).

Regardless of the inability to agree on strategies and co-ordinate action, reef studies mostly show continuing and worrying declines (Burke et al. 2011). In 2007, the IUCN added some 20 coral species to their Red List for the first time in its history (IUCN 2007). This action was vehemently debated among reef researchers, some arguing that this move is nothing more than a political ploy to appropriate territory and only serves to limit access and research opportunities in reef zones, given the complicated permitting and sample-gathering and handling restrictions that Red List status entails. And in 2008, a NOAA report called 'The State of Coral Reef Ecosystems of the United States and Pacific Freely Associated States' stated that slightly over half of the coral reefs in U.S. jurisdiction are in 'poor' or 'fair' condition (Waddell and Clarke 2008).

Although MPAs do not necessarily fall into the realm of natural sciences, ideally they are guided by them. As such, reef scientists have taken on MPA policy as attendant missions, which has deepened debate and contention in terms of 'best fixes' and most important policy-strategies. Scientists, like everyone else, tend to preference their ideas and often prioritize policy strategies in accordance with the sub-topics they work on (Frazzetto 2004, Jain et al. 2009). When seen from a wider perspective that considers all relevant studies, science portrays every reef

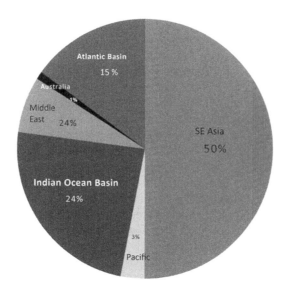

Fig. 4.3   **Distribution, in percent, of the estimated 275 million people who live within 30 kilometres of a coral reef.**

Source: Bischof 2010a.

stressor as a priority to address. Importantly and not as obvious, however, is that most reef scientists do not consider the relevance of people and communities outside of their role as an abusive stressor on reef systems. Political ecological perspectives, although potentially contributing a great deal towards understanding other, 'external' forces that inhibit sustainability of a resource (see Durrenberger and Palsson 1987, Smethurst and Nietschmann 1999) often remain outside the scope of the scientific projects as 'social issues' and in the scientific community, are categorized as being within the purview of policy-makers and enforcement strategies. This has led to suggested policies by scientists such as banning all fishing on reefs (e.g. Jackson et al. 2001), regardless of the reef's importance in subsistence systems of developing countries and the reliance of reefs in poverty-stricken communities, of which there are many in coral reef regions.

Geographies of Science and Uncertainty

Despite nearly two centuries of reef study (Dana 1853, Dobbs 2005, Miner 1925, Vaughan 1910), scientists and managers still cannot seem to agree regarding which problems are most urgent and which management strategies are appropriate often resulting in volatile debate and inaction (Bischof 2010b, Dimitrov 2002, Kleypas and Eakin 2007). Reports of their demise are attributed to an enormous variety of causal factors, all of which relate to both location and socio-economic activities (King 2005, Kleypas and Eakin 2007). Debates regarding environmental management and policy goals for marine conservation are therefore fraught with competing statements that have been sifted through disciplinary screens and geographic filters, and which are applied to both scientific results themselves as well as how they are interpreted, funded, consumed and reproduced (Delaney and Hastie 2007, Kleypas and Eakin 2007, Price 1971).

These contentions of how to prioritize both problems and solutions serve as a serious obstacle, sometimes scientific, sometimes political, but most often both, and always rest on rhetoric of scientific uncertainty and critique (Delaney and Hastie 2007, Dimitrov 2002, Latour 2004, Price 1971). For example, monitoring methodologies are a significant point of controversy as a standard is sought to generalize about reef-health across regions. The lack of agreement among coral reef scientists about which species to count, how to set up transects, and how to count them has led to more than two dozen different reef survey methods and an abundance of data that is not possible to compare or contrast across space or regions, and accompanied by a slew of questionable interpretations of ecosystem form and function and what conditions would qualify a reef to be 'healthy', 'in decline', or 'resilient' given the enormous variation in biological diversity and species types and abundance. (Ginsburg 1994, 2012; IUCN 2007). And, compounding these debates is that although all researchers agree that reefs are in steep decline regardless of survey method applied, causal explanations of environmental destruction are deeply rooted to the geography out of which these data emerged and the researchers' core attitudes regarding human-nature relationships in place. The differences between preferred scientific interpretations and ontological claims about reef systems ecologies show clear geographic trends. Poverty, poaching and excessive exploitation practices are seen as the major problems in Indo-Pacific reef systems, which host the highest biodiversity; and overfishing, invasive species, and physical damage to the reef structure are considered biggest problems in the western Atlantic, with over-population presumed as the driver (Bischof 2010a, Hughes et al. 2002, Kleypas and Eakin 2007, Livingstone 2003) (see also Figure 4.1).

Disagreements in academia regarding actions for environmental systems, including coral reefs, originate from differing attitudes or personal beliefs and are always expressed in terms of scientific uncertainty and epistemological uniquenesses (Bischof 2010a, Latour 1987, 2004). Research involving coral reef scientists and professionals from around the world revealed that the material circumstances inherent to marine spaces, and their mutual understanding of the difficulty of collecting marine field-data, greatly affected the certainty and discourse about circumstances in reef-rich region. Additionally, because of the challenges in working in marine spaces, heuristic elements are often a primary driving force of debate and arguments centered on disparate personal experiences and anecdotes dominate informal discussions and histories of place (Bischof

2010a, Dimitrov 2002, Ginsburg 2012, Kleypas and Eakin 2007, Powell 2007, West and Brockington 2006).

Along with the inherent bias included in the geographic variations of the conditions, features and problems of reefs, uncertainty also lies in the scientific processes of conducting research and developing paradigms in these regions. The processes of conducting research in the marine environment is unique and particular to the material contingencies and tremendous challenges imposed by conducting work in these marine systems, enormous, hazardous (to humans) and difficult (Deacon 1971). The relative few who are able (financially) and willing (in terms of risk and personal challenges inherent when at sea) to engage in these ventures results in distinct epistemologies and a tight social network that although its members are in general agreement about the undeniable plight of reefs globally, argue on the basis of what they have personally seen. Steinberg (2008) has shown that a conflation of the marine environmentalist discourses of 'overuse' and 'underexposure' have a pivotal impact on management philosophies applied in marine space. This duality can be seen in the culture of reef scientists: on one hand, they all agree that the ocean systems are being 'overused' and communicate that concern to all management professionals and reef conservation activists when asked; and on the other hand, in terms of 'underexposure', the relatively small network limits the number of those who have witnessed and tracked the declining trends on individual reef regions and field sites. In narratives among reef scientists there also resonates notions of retrieving the ideal of what a reef should be. Within these professional networks, senior scientists often show or invoke images of times past, when they remember how lush and healthy their field site once was and often apply this nostalgia as a motivating force to find some way to mitigate reef decline, but importantly as an authority to decide the best way to do so, often vilifying or marginalizing the social systems (e.g. subsistence economies, postcolonial development frictions) that, despite the best intentions, they commonly do not or can not take into account as issues and factors that remain outside of their disciplinary domains (e.g. Jackson 1997; Jackson et al. 2001).

Because of our perceptions of physical oceans as a vast frontier, and a space of both separation and connection, inherent access limitations, particular legal conventions, and the variety of technological capabilities required to function in this realm, ocean space plays an influential role in international relations. Water in general has long been seen as a separating force (for example medieval Europeans' use moats around castles and strong-holds; islands as 'isolated', etc.); however for those with resources, oceans present a mechanism for mobility and specialized access and opportunity for control. The combination of socio-political relevance of oceans combined with the mechanisms of science that are active there result in a highly particular and different set of internalized constructs of ocean geographies that directly and concretely impact our behavior and management processes in this space (Bischof 2010b, Kroll 2008, Steinberg 1999, 2001, 2009).

Defining Subjectivity and Revealing Worldviews

Understanding the core beliefs and attitudes that are common in the epistemic community of reef researchers, therefore, has the potential to untangle the social forces that influence narratives and shape discourses regarding reef science and conservation from the agreed truth-claims about reefs, thus separating what is *known* to what is *championed* based on geographies, expertise, and socially-derived specifics that veil common concerns that all can agree with and potentially act on. The resulting confusion between issues that are matters of concern as opposed to matters of fact, in part also because of the manner in which science is communicated and disseminated, can be primarily faulted for what amounts to paralysis in comprehensive postulations of reef ecosystems, debates over dominant ecological driving forces, and the real (versus perceived) issues that have hampered effective conservation efforts (Latour 2004, 1987; Powell 2007). Therefore by underemphasizing the scientific uncertainties that override any consensus on how to mitigate rates of reef declines, and by distilling embedded geographic differences and subjectivities that influence individual preferences, the variety of issues posited as most urgent can be organized in terms of deeper attitudes found in the epistemic network itself and provide explanatory power to decisions that are made regarding actions taken within the ecosystem itself.

By using Q-methodology (Q), a mixed method that applies factor analysis and discourse analysis to find subjective elements that influence the variety of truth-claims and negotiate geographic difference, priorities and levels of urgency are revealed. Q is particularly useful in contentious topics as it is able to prioritize features that are given equal valence. This method was established in 1935 and was first applied in political science and psychology but has gained popularity in a wide range of disciplines (Brown 1980). Q was applied here as a particularly useful way to understand the subjective nature of coral reef conservation and the need to delineate fact from preference that comes from both the social network of reef researchers and the variety of geographies in which their work is conducted.[3]

Topics in reef research that involved the greatest contention based on the issues described above were addressed in this particular study, namely:

1. Geographic Perspectives: a) Global b) Regional/Local c) De-located/Universal
2. Ecological Conditions: a) Ecosystem shift vs extinction b) Location: remote vs proximal Scale of Problem: local vs global.
3. Dominant Threats to Ecosystem: a) global stressors b) local stressors
4. Viewpoints on Human Population: a) overpopulation b) unsustainable development c) scales of influence
5. Overall Attitude: a) global b) local c) degree of certainty

---

3 For primers on Q-methodology and their applicability in human geography, see Brown (1980), Eden et al. (2005), McKeown and Thomas (1988), and Robbins and Krueger (1999).

Results and Conclusions: Finding Trends of Consensus and Difference

The core attitudes among coral reef scientists are divided in three distinct worldviews regarding the way in which they imagine ocean systems in terms of issues examined and are described in Table 4.1. Most noticeably, a pervasive trend that explains the dominant discourses of environmental decline found in all branches of marine ecosystems researches is rooted in an overwhelmingly Malthusian position among scientists in the construction of marine environmental crisis (Bischof 2010a, Pelletier et al. 2005, Souter and Linden 2000). This position, along with the manner in which scientists conceive human-environment interaction as revealed in the table below (Table 4.1), supports the notion that because oceans are encountered as both objects of consumption and objects of aesthetic utility, the role of humans in the environment becomes confusing and ambiguous, and perspectives that idealize imaginations of ocean systems, by default, limit the role of humans in the environment to that of destructive agents (Jackson and Sala 2001, King 2005).

Determining these features can inform policy measures that roll back the influence of individual preference by quantifying subjectivities of the issues. This permits an ordering of priorities based on analysis of the entire epistemic network and extracts, for the most part, assumptions based on individual influence, power, social constructions and imaginations of reef space as an 'expert', or the prestige of institutional affiliations. When conducting the Q-study, distinctive attitudes can be described based on how the respondents organized the statements they sorted in the exercise. The dominant attitudes (the number of those who fell into those particular viewpoints) are Gain Communalists, followed by Science-oriented Communalists, and with the fewest sorting into the Locally-oriented Positivists grouping. These were the three main 'attitudes' that emerged, and were based in how statements were prioritized. Gain Communalists were most concerned with how the smaller regions fit together and constructed a big picture of reef issues; science-oriented communalists were far more concerned with how to integrate scientific findings, and focused on the empirical aspects as those that should direct behavior; lastly, locally-oriented positivists were most interested in how change can be brought about through community and/or local involvement.

From Table 4.1, it becomes clear that differences among those involved in reef conservation can be distinguished by geographic perspectives, ecological conditions, dominant threats and perceptions of human systems and align with distinct core philosophical viewpoints. Understanding how scientists who have influences on policy measures and ways towards their implementation perceive, in this case, reef environments allows significant and necessary insight into their biases and can separate those issues that are more generally agreed to be important versus importance based on personal preferences. Because the perception of environment plays the largest role in how we respond to its alterations, whether natural [sic] or anthropogenic, it is crucial to first determine what those perceptions may be based on in terms of how environments are constructed out of preferences of fact.

Table 4.1   The dominant attitudes revealed by the Q-study factor analysis.

	Gain Communalists	Science-oriented Communalists	Locally-oriented Positivists
Geographic Perspectives	• 'Global' and 'holistic' vision of the environment as part of a larger system, but made up of smaller regions that are known through locally-relevant and scientifically-based information. • Local issues are important, but only as far as being an explanatory part of the global issues.	• Objective and academically-situated view of the human-environment feedback system. • Society should respond in accordance to the available scientific knowledge as the foundation of management strategies. • Sensibilities of geographic difference between development status of countries, however these differences do not change scientific understandings.	• Highly critical of local abuses of reefs, although seen as a problem of human population densities. • Global changes manifest in local or regional scales • Reliable observations of reef environments happen on local and regional scales. • Little importance given to political and economic differences, but community viewpoints and use-regimes most relevant.
Ecological Condition	• The ecosystems are experiencing a shift, not an extinction. • The system can be handled as one that is adaptive and mutable. • Reefs are locally distinct but only insofar as representing part of a global ecological network of reef characteristics and as such, little emphasis is placed on local stressors but rather how these changes represent a cumulative response of the global reefs.	• Isolated reefs are in the best condition and biodiversity is of utmost importance. • Consequences of increased CO2 and ocean warming are urgent and of primary concern in the long-term future. • Coral ecosystems are experiencing an extinction event. • The current situation shows no clear signs of recovery nor do scientific results indicate any optimism about the future, reinforcing science as the authoritative source.	• Connectivity between reef systems should be sought rather than focusing on concerns over biodiversity and resilience. • Local problems are most important such as local trends of the curio/aquarium trade, overfishing, destructive tourism practices, deforestation, coastal development, invasive species, anchor damage, ghost traps, etc. • Isolated reefs are in far better condition than those near land and population centers.
Dominant Threats	• Concern about global stressors together are most important and the biggest concern, including warming oceans but also the coral disease. • Stressors are a result of the human communities using reefs and are inevitable as populations increase both in general and along the coasts.	• Ocean acidification and increasing levels of CO2 are the most immediate threats; local stressors such as the curio trade, invasive species, localized physical damage will determine how global threats will emerge in each environment. • Human abuses and unsustainable exploitation	• Human populations increasing exponentially, both along the coasts and in general. • Coastal development • Expansion of commercialized, unsustainable resource extraction, especially fisheries. • Localized destructive practices such as deforestation, irresponsible tourism practices, excessive local resource extraction.

	Gain Communalists	Science-oriented Communalists	Locally-oriented Positivists
Viewpoint on Population	• The ecosystems are experiencing a shift, not an extinction. • The system can be handled as one that is adaptive and mutable. • Reefs are locally distinct but only insofar as representing part of a global ecological network of reef characteristics and as such, little emphasis is placed on local stressors but rather how these changes represent a cumulative response of the global reefs.	• People will inevitably behave badly and unsustainably exploit and abuse reef environment as long as resources are available. • The problems in reef-rich areas have to be resolved through stronger enforcement and management oversight. • It is not community populations near the reef sites that should be faulted for reef conditions, but the non-native population and associated resource extraction pressures resulting from greater population densities in coastal regions.	• Rising populations are the most urgent issue to be dealt with on all scales, although subsistence or local use-regimes are not the main problem, but rather commercial industries. • Important that the local and subsistence community needs are understood if conservation efforts are to be effective. • Population is the dependent variable from which destructive environmental causation can be determined
Overall Attitude	• Holistic 'all in it together'. • Community is responsible as a matrix of the global system. • 'Think global, act local' attitude; strong faith in the power of community.	• Distant and separated view from the needs of communities reliant on reef populations. • Cynical about the future and human potential for positive change. • Emphasis on global problems, especially global CO_2 fluctuations and resulting ocean acidification	• Strong faith in scientific discourses of reef decline, particularly as how they can be integrated into community behavioral responses • Emphasis on the local (geographically) insofar as each region should focus on their specific problem/issues rather than focus on global concerns such as warming and acidification.

Source: Bischof (2010a)

Although all reef researchers, both those who study social systems as well as physical systems, agree that coral reef ecosystems are in a tail-spin of decline and require immediate attention, their views of problems and solutions to halt or reverse that decline are fixed deeply within the particular attitudes out of which they imagine ocean systems, geographies of their field sites, and the role of humans in environment. The polarization of these views within a community that is sought to provide answers for environmental management therefore plays a major role in deepening competing ideologies and, in effect, results in impotent management structures that are rooted in the foundational differences in how coral reef researchers imagine ocean space on global and local scales and which actions they suggest as ways of mitigating environmental stressors.

It is generally agreed among scientists themselves that the volume of knowledge about this rapidly-changing system is adequate to generate cohesive policy directions on all scales, the deep-seated polarized beliefs and attitudes within the epistemic community of reef-researchers are an enormous obstacle that serves to hamper consolidated efforts of environmental management in these extremely complex systems. But by rendering the underlying forces within the social networks of those defining the ontological premises for environmental conservation of a globally-distributed ecological system, this work provides evidence that a geographic perspective offers decisive insights that reach beyond the primarily reductionist strategies of the natural sciences.

Replacing the embedded features of geographic difference in de-located scientific discourses by outlining the core attitudes held within the epistemic community, reveals not only the collective perceptions of people in place as a defining element of environmental condition of global reefs, but also shows that the particular preferences of the researchers themselves greatly influence how this relationship is conceived and judged, and addresses the limitations of natural scientific inquiry alone in advancing environmental conservation in an increasingly troubled ecosystem.

References

Anthony, K.R.N., D.I. Kline, G. Diaz-Pulido, S. Dove and O. Hoegh-Guldberg. 2008. Ocean acidification causes bleaching and productivity loss in coral reef builders. *Proceedings of the National Academy of Sciences of the United States of America*, 105(45), 17442–6.

Baker, A.C., Glynn, P.W., and Riegl, B. 2008. Climate change and coral reef bleaching: An ecological assessment of long-term impacts, recovery trends and future outlook. *Estuarine, Coastal and Shelf Science*, 80, 435–71.

Birkeland, C. 1997. *Life and Death of Coral Reefs*. Springer: New York.

Bischof, B. 2010a. The Coral Reef Environmental "Crisis": Negotiating Knowledge in Scientific Uncertainty and Geographic Difference. Florida State University, Department of Geography, Dissertation. May 2010.

Bischof, B. 2010b. Negotiating uncertainty: Framing attitudes, prioritizing issues, and finding consensus in the coral reef environment management 'crisis'. *Ocean & Coastal Management*, 53, 597–14.

Bowler, P. 1992. *The Norton History of Environmental Sciences*. W.W. Norton & Company: New York.

Brown, S. 1980. *Political Subjectivity: Applications of Q-Methodology in Political Science*. New Haven: Yale University Press.

Burke, L.K., Reytar, M. and Spalding, and Perry, A. 2011. *Reefs at Risk Revisited*. World Resources Institute: Washington DC.

Caddy, J.F. and Garibaldi, L. 2000. Apparent changes in the trophic composition of world marine harvests: The perspective from the FAO capture database. *Ocean & Coastal Management*, 43, 615–55.

Carr, M.H., J.E. Neigel, J.A. Estes, Andelman, S., Warner, R.R., and Largier J.L. 2003. Comparing Marine and Terrestrial Ecosystems: Implications for the Design of Coastal Marine Reserves. *Ecological Applications*, 13(1), S90–S107.

Cowen, R.K., Paris, C.B. and Srinivasan, A. 2006. Scaling of connectivity in marine populations. *Science*, 311, 522–27.

Dana, J.D. 1853. *Coral Reefs and Islands*. New York: G.P. Putnam.

Deacon, M. 1971. *Scientists and the Sea, 1650–1900: A Study of Marine Science*. London: Academic Press.

Delaney, A.E., and Hastie, J.E. 2007. Lost in translation: Differences in role identities between fisheries scientists and managers. *Ocean & Coastal Management*, 50, 661–82.

Dimitrov, R. 2002. Confronting nonregimes: Science and international coral reef policy. *The Journal of Environment & Development*, 11(1), 53–78.

Dobbs, D. 2005. *Reef Madness: Charles Darwin, Alexander Agassiz, and the Meaning of Coral*. New York: Pantheon.

Durrenberger, E.P. and Palsson, G. 1987. Ownership at sea: Fishing territories and access to sea resources. *American Ethnologist*, 14(3), 508–22.

Eden, S., Donaldson, A. and Walker, G. 2005. Structuring subjectivities? Using Q methodology in human geography. *Area* 37(4), 413–22.

Frazzetto, G. 2004. The changing identity of the scientist. *European Molecular Biology Organization Reports*, 5(1), 18–20.

Ginsburg, R.N. 2012. Emeritus Professor, University of Miami Rosenstiel School of Marine and Atmospheric Science, Division of Marine Geology and Geophysics. Personal communication.

Ginsburg, R.N. (ed.). 1994. *Proceedings of the Colloquium on Global Aspects of Coral Reefs: Health, Hazards, and History 1993*. Rosenstiel School of Marine and Atmospheric Science, University of Miami. Miami, FL.

Guzman, H.M., Guervara, C., and Castillo, A. 2003. Natural disturbances and mining of Panamanian coral reefs by indigenous people. *Conservation Biology*, 17(5), 1396–401.

Hargreaves-Allen, V., Mourato, S., and Milner-Gulland, E.J. 2011. A global evaluation of coral reef management performance: Are MPAs producing conservation and socio-economic improvements? *Environmental Management*, 47, 684–700.

Hughes, T.P., Baird, A.H., Bellwood, D.R., Card, M., Connolly, S.R., Folke, C., Grosberg, R., Hoegh-Guldberg, O., Jackson, J.B.C., Kleypas, J., Lough, J.M., Marshall, P., Nyström, M., Palumbi, S.R., Rosen, B. and Roughgarden, J. 2003. Climate change, human impacts, and the resilience of coral reefs. *Science*, 301, 929–33

Hughes, T.P., Bellwood D.R. and Connolly S.R. 2002. Biodiversity hotspots, centers of endemicity, and the conservation of coral reefs. *Ecology Letters*, 5, 775–84.

IUCN. 2007. *Extinction Crisis Escalates: Red List shows apes, corals, vultures, dolphins all in danger* [Online] Available at: http://iucn.org/media/news_ releases/?81/Extinction-crisis-escalates-Red-List-shows-apes-corals-vultures-dolphins-all-in-danger. [accessed 14 June 2012].

Jackson, J.B. 1997. Reefs since Columbus, *Coral Reefs*, 16, Suppl. S23-S32.

Jackson, J.B., Kirby, M.X, Berger, W.H., Bjorndal, K.A., Botsford, L.W., Bourqu, B., Bradbury, R.H., Cooke, R., Erlandson, J., Estes, J.A., Hughes, T.P., Kidwell, S., Lange, C.B., Lenihan, H.B., Pandolfi, J.M., Peterson, C.H., Steneck, R.S., Tegner, M.J. and Warner, R.R. 2001. Historical overfishing and recent collapse of coastal ecosystems. *Science*, 293(5530), 629–37.

Jackson, J.B. and Sala, E. 2001. What was natural in the coastal oceans? *Proceedings of the National Academy of Sciences*, 98(10), 5411–18.

Jain, S., George, G. and Maltarich, M. 2009. Academics or entrepreneurs? Investigating role identity modification of university scientists involved in commercialization activity. *Research Policy*, 38, 922–35.

Jentoft, S., Van Son, T.C. and M. Bjørkan, M. 2007. Marine protected areas: A governance system analysis. *Human Ecology*, 35(5), 611–22.

Juda, L. and Hennessey, T. 2001. Governance profiles and the management of the eses of large marine ecosystems. *Ocean Development & International Law*, 32, 43–69.

Kenchington, R. 1998. Status of the international coral reef initiative. In *Coral Reefs: Challenges and Opportunities for Sustainable Management*, edited by M.E. Hatziolos, A.J. Hooten and M. Fodor. World Bank: Washington, D.C., 11–15

King, T.J. 2005. Crisis of meanings: Divergent experiences and perceptions of the marine environment in Victoria, Australia. *The Australian Journal of Anthropology*, 16(3), 350–65.

Kleypas, J.A. and Eakin, C.M. 2007. Scientists' perceptions of threats to coral reefs: Results of a survey of coral reef researchers. *Bulletin of Marine Science*, 80(2), 419–36.

Kleypas, J.A. and Yates, K.K. 2009. Coral reefs and ocean acidification. *Oceanography*, 22(4), 109–16.

Knowlton, N. 2001. The future of coral reefs. *Proceedings of the National Academy of Sciences*, 98(10) 5419–25.

Knowlton, N. 2008. Coral reefs. *Current Biology*, 18(1), R18–R21.

Kroll, G. 2008. *America's Ocean Wilderness: A Cultural History of Twentieth-Century Exploration.* Lawrence: University Press of Kansas.

Latour, B. 1987. *Science in Action: How to Follow Scientists and Engineers through Society.* Cambridge: Cambridge University Press

Latour, B. 2004. Why has critique run out of steam? From matters of fact to matters of concern. *Critical Inquiry*, 30, 225–48.

Livingstone, D. 2003. *Putting Science in its Place: Geographies of Scientific Knowledge.* Chicago: University of Chicago Press.

Longhurst, A. 2003. The symbolism of large marine ecosystems. *Fisheries Research*, 61, 1–6.

McLanahan, T.R. 2011. Coral reef fish communities in management systems with unregulated fishing and small fisheries closures compared with lightly-fished reefs—Maldives vs Kenya. *Aquatic Conservation: Marine and Freshwater Ecosystems*, 21(2), 186–98.

McKeown, B. and Thomas, D. 1988. *Q Methodology.* Series: Quantitative Application in the Social Sciences, No. 66. London: Sage.

Miner, R.W. 1925. Reef Builders of the Tropic Seas. *Natural History*, 25(3), May–June, 250–60.

Mora, C., Aburto-Oropeza, O., Bocos, A., Ayotte, P.M., Banks, S., Bauman, A.G., Beger, M., Bessudo, S., Booth, D.J., Brokovich, E., Brooks, A., Chabanet, P., Cinner, J.E., Cortes, J., Cruz-Motta, J.J., Cupul-Magaña, A., DeMartini, E.E., Edgar, G.J., Feary, A., Ferse, S.C.A., Friedlander, A.M., Gaston, K.J., Gough, C., Graham N.A.J., Green, A., Guzman, H., Hardt, M., Kulbicki, M., Letourneur, Y., Lopez-Perez, A., Loreau, M., Loya, Y., Martinez, C., Mascarenas-Osorio, I., Morove, T., Nadon, M-O., Nakamura, Y., Paredes, G., Polunin, N.V.C., Pratchett, M.S., Bonilla, H.R., Rivera, F., Sala, E., Sandin, S.A., Soler, G., Stuart-Smith, R., Tessier, E., Tittensor, D.P., Tupper, M., Usseglio, P., Vigliola, L., Wantiez, L., Williams, I.,. Wilson, S.K. and Zapata, F.A. 2011. Global human footprint on the linkage between biodiversity and ecosystem functioning in reef fishes. *PLoS Biology*, 9(4), DOI: 10.1371/journal.pbio.1000606.

Mora, C., Andrefouet, S., Costeloo, M.J., Kranenburg, C., Rollo, A., Veron, J., Gason, K.J. and Myers R.A. 2006. Coral reefs and the global network of marine protected areas. *Science*, 312, 1750–51.

Orr, D. 1992. *Ecological Literacy.* Albany: State University of New York Press.

Pelletier, D., Garcia-Charton, J.A., Ferraris J., David G., Thebaud, O., Letourneur, Y., Claudet, J., Armand, M., Kulbicki, M. and Galzin R. 2005. Designing indicators for assessing the effects of marine protected areas on coral reef ecosystems: A multidisciplinary standpoint. *Aquatic Living Resources*, 18, 15–33.

Powell, R. 2007. Geographies of science: Histories, localities, practices, futures. *Progress in Human Geography*, 31(3), 309–29.

Price, D.J. 1971. *Little Science Big Science.* New York: Columbia University Press.

Reakla-Kudla, M.L. 1997. The global biodiversity of coral reefs: A comparison with rain forests. In *Biodiversity II: Understanding and Protecting Our Natural Resources*, edited by M.L. Reakla-Kudla, Wilson, D.E. and Wilson, E.O. Washington DC: Joseph Henry 83–103.

Riegel, B.A., Bruckner, S.L., Coles, P., Renaud. and Dodge, R. 2009. Threats and conservation in an era of global change. *The Year in Ecology and Conservation Biology 2009: Annals of the N.Y. Academy of Sciences*, 1162, 136–86.

Robbins, P. and Krueger, R. 1999. Beyond bias? The promise and limits of Q method in human geography. *Professional Geographer*, 52(4), 636–48.

Roberts, C.M., McClean, C.J., Vernon, J.E.N., Hawkins, J.P., Allen, G.R., McAllister, D.E., Mittermeier, C.G., Schueler, F.W., Spalding, M., Wells, F., Vynne, C. and Werner, T.B. 2002. Marine biodiversity hotspots and conservation priorities for tropical reefs. *Science*, 295, 1280–84

Sherman, K. 1995. Achieving regional cooperation in the management of marine ecosystems: The use of large marine ecosystem approach. *Ocean & Coastal Management*, 29(1–3), 165–85.

Smethurst, D. and Nietschmann, B. 1999. The distribution of manatees (*Trichechus manatus*) in the coastal waterways of Tortuguero, Costa Rica. *Biological Conservation*, 89(3), 267–74.

Souter, D.W. and Linden, O. 2000. The health and future of coral reef systems. *Ocean & Coastal Management*, 43, 657–88.

Spalding, M.D., Ravilious, C. and Green, E.P. 2001. *World Atlas of Coral Reefs*. Berkeley: University of California Press.

Steinberg, P.E. 1999. The maritime mystique: Sustainable development, capital mobility, and nostalgia in the world ocean. *Environment and Planning D: Society & Space*, 17, 403–26.

Steinberg, P.E. 2001. *The Social Construction of the Ocean.* Cambridge: Cambridge University Press.

Steinberg, P.E. 2008. It's so easy being green: Overuse, underexposure, and the marine environmentalist consensus. *Geography Compass*, 2(6), 2080–96.

Steinberg, P.E. 2009. Sovereignty, territory and the mapping of mobility: A view from the outside. *Annals of the Association of American Geographers*, 99(3), 467–95.

Stoddard, D.R. 1986. *On Geography and its History*. Oxford: Blackwell.

Vaughan, T.W. 1910. A contribution to geologic history of the Floridian Plateau. *Papers from the Tortugas Laboratory*, Carnegie Institute of Washington Pub. 133, 99–185.

Veron, J.E.N., Bellwood, D.R., and Hughes, T.P. 2001. The state of coral reef science. *Science*, 293, 1996–7.

Waddell, J.E. and A.M. Clarke, A.M. 2008. (eds.) *The State of Coral Reef Ecosystems of the United States and Pacific Freely Associated States: 2008*. NOAA Technical Memorandum NOS NCCOS 73. NOAA/NCCOS Center for Coastal Monitoring and Assessment's Biogeography Team. Silver Spring, MD. 569.

West, P., and Brockington, D. 2006. An anthropological perspective on some unexpected consequences of protected areas. *Conservation Biology*, 20(3), 609–16.

Wilkinson, C. (eds.). 2008. *Status of Coral Reefs of the World 2008*. Global Coral Reef Monitoring Network and Reef Rainforest Research Center, Townsville, Australia.

Chapter 5
Merging with the Medium?
Knowing the Place of the Surfed Wave

Jon Anderson

Introduction

The way human geographers approach the world is changing. How we understand places and spaces, be they terrestrial or aquatic, is shifting from an assumption of ontological fixedness and stability, towards an acceptance of ontological *in*stability. Some scholars go further to suggest that places should now be conceived as constituted by merging and emerging ontologies (see for example, Latour 1993, Dovey 2010; and in respect to post-nature Braun 2004, Hayles 1999). There is not the space in this chapter for an extended discussion of these wider debates (for this, see Anderson 2009), but rather I wish to introduce the ways in which the surfed wave can be known as a place that is at once unstable, emergent, and co-constituted by both 'human' and 'natural' components.

As Shields describes, 'surfing is the art of standing and riding on a board propelled by breaking waves' (2004: 45), or as Ford and Brown put it, 'the core of surfing has always simply been the embodied, raw and immediate glide or slide along a wave of energy passing through water' (2006: 149). Surfing can be practiced in many ways, including long-, short-, or body-boards, on surf-kayaks, or surf-skis. This chapter will focus on those who conjoin the place of the surfed wave through board riding. Drawing on commentaries from surfers on the practice of wave-riding, the chapter will show that the surfed wave can be known in two ways: as an 'assemblage' (see Delanda 1996), and as a 'convergence' (see Anderson 2009). Whilst the notion of assemblage suggests that surfers, boards and waves are 'connected' together to form one coherent unit for the lifetime of the ride, convergence argues that the surfed wave becomes a place whose constituent parts are not simply connected together, rather their thresholds are blurred into a converged entity/process. Theorizing to and from the sea in this way demonstrates how human geographers can begin to consider traditional places in new ways, but also consider new (watery) 'coming-togethers' as 'places'.

The initial research for this chapter was funded by Sports Council for Wales in order to investigate participation regimes in surfing in the principality. The research is based on an online questionnaire which was completed by 134 surfers in Wales; interviews with the proprietors of 14 surf schools in Wales; interviews with 6 surf club secretaries; and 20 in-depth interviews with people who surf in Wales. It is supplemented by extensive secondary accounts of surfing worldwide taken from magazines (including *The Surfers Path*, *Surfer*, *Carve*, *Drift* and *Wavelength*), as well as books, autobiographies, and biographies of surf culture. The author is a (kayak) surfer (see Chapter 7 this volume).

Relational Places

Enabled by the rise of postmodern cultural and social theory (after Oakes 1997: 509), geographers have disrupted and challenged the stable, coherent, and static approach to geographical sites favoured by sedentary metaphysics (see Malkki 1992, Cresswell 2004). The notion of place has changed from one that is sedentary and stable to one that is, or many that are, provisional and emergent. These loosely aligned approaches are often disparate in their specifics (see, for example, the geographies of Doel 1999 and Murdoch 2006; or the actor networks of Callon 1986 and Latour 1999), but as Jones identifies, all approach the notion of place in a *relational* way;

> recent years have witnessed a burgeoning work on 'thinking space relationally'. According to its advocates, relational thinking challenges human geography by insisting on an open-ended, mobile, networked, and actor-centred geographic becoming. (Jones 2009: 5)

Relational thinking marks a shift away from the independent conceptual categories of a sedentary metaphysics and the modern constitution more broadly (see Latour 1993). No longer do the 'noun chunks' of this constitution hold sway (after Laurier and Philo 1999), the fixed and essential notions of 'physical', 'human', 'culture', or 'economy' and their associated binary framings (e.g. 'A – not A') give way to an *inter*dependent epistemology where things are always acting and being acted on by everything else. In Doel's words, these theories approach the world as if it is 'a verb rather than a noun'; rather than something fixed, it is something in the making (2000: 125). These amodern approaches thus opt to focus on the relations and interactions that continually form the world, on how no-longer-isolated categories are bound together in networks or 'relational complexes' (Rouse 1996). This thinking marks a rejection of a static ontology of 'being-in-the-world' (associated with Heidegger 1956, for example), and an embracing of a more emergent and emerging ontology of 'becoming-in-the-world' (associated with Deleuze 1985, 1993).

Thus part of this move towards an emerging ontology includes our relations 'in-the-world'. Acknowledging this intimacy and connection between people and place has led to a reframing of human relations away from traditional scientific notions of detachment from the world, towards recognition of our involvement and interdependence with the world. As a consequence, places are no longer 'presumed to be relatively fixed, given, and separate. ... Rather, there is a complex relationality of places and persons connected through performances' (Murdoch 2006: 214).

There is an acknowledgement that our practices and performances affect places, and in turn places and practices affect us. Allied to these shifts, there is a growing recognition that emotions are integral to our emerging understanding of places. As Vitterso et al. state: 'As a new century begins, attention to affect and emotion has been recognized as an essential impact of any study of humankind' (2001: 137). Emotions, therefore, are increasingly seen as essential components in our knowledge systems, as well as inevitable productions from our interactions with the world of which we are a part; it is through emotions that we 'literally make sense of the world' (Wood and Smith 2004: 534).

From this perspective, places produce relational sensibilities (or emotional reactions that are generated within humans as a result of being part of these relations, see Anderson, 2009). How places are understood are thus dependent on these practices, and the relational affects they produce. Places are no longer definitive and fixed (we no longer see static towns, buildings, countries, or parks), but are actor- and practice-centred becomings – we have ski-ed hillsides, walked streets, taught seminar halls, slept doorways, etc. Places can be plural and provisional; place is no longer sedentary and stable, but evolving and emergent.

The Surfed Wave as a Relational Place

> Surfing was the thing to do, and this was the *place* to be. (Kampion 2004:20, emphasis added)

It is in this context that this chapter argues the 'surfed wave' can be understood as a relational place. As Hill and Abbott identify, surfing, let alone the surfed wave in particular, is not generally considered an accepted focus of academic study (2009: 276). Notable exceptions exist, however, including the excellent overview of surfing by Stranger (2011), as well studies by Nazer (2004), Ford and Brown (2006) and Evers (2006, 2009). Surf itself is clearly both a fluid and temporary phenomenon, and as such would not necessarily be considered as a place from a sedentary, terrestrial perspective. The location of the beach, or the place where surf is likely to occur (for example, an established reef or surf break, see SurfingBlog 2009) is more commonly studied as a place (see Shields 2004) – these places are where surf is (more) predictable and located, it is where the transience and fluidity of waves have a degree of permanence. However, I would like to argue here that a specific surfed wave (as opposed to a surf break, or even set of waves) can also be considered as a place from a relational perspective. A single wave may be difficult to predict, let alone locate, as its existence depends on a range of factors, as Raban points out: 'a single wave is likely to be moulded by several forces: the local wind; a dominant, underlying swell; and, often, a weaker swell coming from a third direction' (1999: 93). In the same vein, Shields notes that the presence of rideable surf is 'highly dependent on the nature and pattern of incoming waves – their height, undertow, whether they break parallel or at an angle to a beach, and so on' (2004: 45). Thus for the place of surf to actually exist requires the interdependent and provisional coming together of a range of components – sea, swell, wind, continental shelf, reef, tide, etc. At its essence therefore the place of surf is never stabilized or normalized, but conditional on the intersection of a range of changing factors. Due to the fluid and mobile nature of a wave it is, 'always in a process of dynamic unfolding and becoming' (Rose 2002: 385). The place of surf is the very definition of a place that is unreliable, inconsistent, wholly provisional and unstable. It is a place that, at any moment, emerges in time and space from the web of flows and connections meeting at a particular node.

These factors have encouraged both scholars and surfers to consider the phenomenon of surf in relational terms. Sheller and Urry, for example, note that the general location of surf (where the beach and the sea meet) should not be considered a unitary object, but a 'complex system of diverse intersecting mobilities' (2004: 6). Surf-writer Sean Doherty (2007) describes the creation of surf as a, 'magic amalgam of rock, sand and saltwater'. Here, the place of surf is described not simply in terms of it 'final' state – as a rideable wave – but in terms of its constituent origins and their storied trajectories; this amalgam is 'magic' and mobile in terms of its components' histories and the unlikely coincidence that their trajectories intersect to create this particular place. Yet for the 'surfed wave' to exist, these components also need to be conjoined with a surfer. The surfed wave thus only emerges through practice; it is through the act of surfing that the surfed wave exists. Yet this coming together produces meanings, (re)presentations, and emotional affects that outlive its own existence. The emotions formed as constituent part as well as resulting product of the surfed wave are commonly expressed as 'stoke' (see Kampion 1997). As Stranger (2011) describes, being 'stoked' describes the high the thrill of surfing can induce, it refers to the 'ecstatic reaction to a surfing session or a particular ride' (Stranger 2009: 275). As one surfer puts it:

> The release [from the act of surfing] is incredible. I've heard it from everybody else. You can see the goosebumps on their bodies when they talk about it. … There is no drug like it. There is nothing [else] on earth that has done that to me.
> (surfer in Lyon and Lyon 1997: 140, cited in Stranger 2011: 120)

Each surfer will experience and define 'stoke' differently, but by paying attention to these key differences the surfed wave can be defined in two key ways: as 'assemblage' and as 'convergence'.

'Flow' Experience and the Surfed Wave

For many wave riders the stoke experienced when being involved with the place of the surfed wave can be theorized as 'flow' experience (see, for example, Shields 2004:50, Stranger 1999; and Ford and Brown 2006). This 'flow' experience leads to understanding the surfed wave as assemblage. The 'flow experience' is defined by

Csíkszentmihályi (1990: xi) as a the joy and creativity registered through a 'process of total involvement with life'. It is 'optimal state of experience in which an individual feels cognitively efficient, deeply involved, and highly motivated with a high level of enjoyment' (Asakawa 2009: 123). To experience flow is to have a sense that,

> one's skills are adequate to cope with the challenges at hand, in a goal-directed, rule-bound action system that provides clear clues as to how well one is performing. Concentration is so intense that there is no attention left over to think about anything irrelevant, or to worry about problems. Self-consciousness disappears and the sense of time becomes distorted. (Csíkszentmihályi 1990: 71)

The flow experience is thus focused on the practice, execution, and improvement of skills in order to successfully meet the challenges set, with concentration so focused that other issues, or even a sense of time, no longer seem important. Flow is often experienced by surfers due to the skills necessary to 'catch' a wave and thus ride it. Many surfers express experiences similar to flow, as the following anonymized respondents recounted in interview:

> I wouldn't say that I am hunting for that most awesome wave – I am just happy if I can get up and manoeuvre my board and feel like I am sort of controlling my ride on the wave if you like – I am content with that …

> The reason why I carry on surfing is just the pure challenge, you know that you can ride waves from 2 foot to 10 foot and everyone is different and also you want to get better at it – it is a sport that you are doing with nature and everything like that and you just really do want to get better at it all the time.

> when you're out there and when you surf you've got to focus on what you're doing so you're in that time and place right there, you're there. There's nothing else going on in your mind at the time. Whereas when you're at home or you're at work you can still have four or five different things going on in your head like multitasking and stuff but when you surf you can only have that one thought in your head, and that is what you're doing right at that moment. I think that's probably why.

From being part of the surfed wave surfers experience the sensation of 'flow'. At these moments the place of the surfed wave is defined by their total concentration and mental immersion in the task at hand; the wave demands their immediate and present involvement. The challenge presented by catching the wave and contributing a particular practice to it (e.g. a specific cutback or manoeuvre) means new flow sensations define each surfed wave. The surfed wave is defined by the transient, provisional, and interdependence of each coming together, alongside the 'flow' experiences produced by them. I would argue that this 'flow experience' is thus not only an outcome of being part of the surfed wave, but also goes some way to defining it, in this case as an 'assemblage'.

Introducing Assemblage

As Phillips (2006) outlines, the notion of the 'assemblage' has gained academic currency as a key way to conceptualize relational (terrestrial) places (see also Anderson and McFarlane 2011). The assemblage displays many of the key aspects of an amodern approach: it is relational, provisional, and interdependent in its formation, as Dovey (2010: 16) suggests:

> In the most general sense an 'assemblage' is a whole 'whose properties emerge from the interactions between parts' (Delanda 2006: 5). … The parts of an assemblage are contingent rather than necessary, they are aggregated … ; as in a 'machine' they can be taken out and used in other assemblages. (Dovey 2010: 16)

An assemblage is therefore a component that is formed by the coming together of many other parts. These parts do not come together necessarily by intention or design, or have an essential permanence that makes their connection insoluble, rather their aggregation keeps their individual unitary coherence intact, but nevertheless forms a larger whole through their connection with others. The notion of the assemblage is inspired by Deleuze and Guattari (1981). This chapter is not the place for a detailed exposition of the rudimentary theory of assemblages put forward by Deleuze and Guattari, and then expanded by Delanda (but see Phillips 2006, Marcus and Saka, 2006, Anderson and McFarlane 2011), suffice it to say that, according to Delanda,

> A theory of assemblages, and of the processes that create and stabilize their historical identity, was created by the philosopher Gilles Deleuze in the last decades of the twentieth century. This theory was meant to apply to a wide variety of wholes constructed from heterogeneous parts. Entities ranging from atoms and molecules to biological organisms, species and ecosystems may be usefully treated as assemblages. (2006: 3)[1]

In terms of the scope of this chapter, Dovey has directly applied the notion of the assemblage to the geographical imagination, arguing for the 'conception of place as a territorialized assemblage' (2010: 17). As he explains: for example,

> a street is not a thing nor is it just … a collection of discrete things. The buildings, trees, cars, sidewalks, goods, people, signs, etc. all come together to become the

---

1   As Dovey puts it, for Deleuze the assemblage is 'a kind of hinge for what he terms "assemblage theory". Philosophically this is an attempt to avoid all forms of reductionism – both the reduction to essences and reduction to text. It is empirical without the essentialism of empirical science; it gives priority to experience and sensation without the idealism of phenomenology; and it seeks to understand the social construction of reality without reduction to discourse' (Dovey 2010: 16).

street, but it is the connections between them that makes it an assemblage or a place. It is the relations of buildings-sidewalk-roadway; the flows of traffic, people and goods; the inter-connections of public to private space, and of this street to the city, that make it a 'street' and distinguish it from other place assemblages such as parks, plazas, freeways, shopping malls and marketplaces. Within this assemblage the sidewalk is nothing more than a further assemblage of connections between things and practices. The assemblage is also dynamic – trees and people grow and die, buildings are constructed and demolished. It is the flows of life, traffic, goods and money that give the street its intensity and its sense of place. All places are assemblages. (2010: 16)

For Dovey, (terrestrial) places can be considered assemblages as they are not discrete, essential things in themselves, but rather are formed by the coming together of more or less random component parts. It is how these component parts relate, connect and interact that forms particular places. These relations are not permanent, and therefore places may change, decay, and transform. Similarly, component parts can become detached or removed, and the place become something else. Places as assemblages are not simply constituted by 'things', but also practices (e.g. the building of houses or the movement of people). As a consequence, these places are also constituted by the experiences of those involved in these practices, the meanings they bring to these places, and the intensities of affect produced by their interactions with the other connecting parts. As Dovey suggests, 'change any of these and [the street] would still be a place, but not the same place' (2010: 24).

From this perspective, therefore, places are assembled (with differing degrees of intention); they are assemblies of a range of component parts. Like a child's toy-world, places are connected together, part by part (or 'brick' by 'brick'). As places become assembled, the connecting 'bricks' can still be identified, removed, or become part of something else, but through their assembly also come together to form something larger; the territorialized place itself.[2]

The Surfed Wave as Assemblage

When territorialized into a fixed, locatable site such as the street (as Dovey outlines), the notion of the assemblage is used to demonstrate the interdependent and transitory nature of places traditionally considered as discrete, durable, and permanent. However, this notion can also be used to understand more obviously fluid and temporary places, such as the surfed wave. Taking the territorialized notion of assemblage to the sea has to date been undertaken by Ford and Brown

---

2 Assemblages may also be compared to 3-D scientific models used to create molecular forms. In these each element or atom can at once be identified, removed, or conjoined, retaining its identity as well as forming a larger composite molecule.

(2006), theorizing the surfing body as assemblage. In their insightful account of the ways in which both the surfer and surfing can be theorized, the surfing body can be seen to involve the interrelation of 'genetics, neurophysiology, tools (surfboard, wetsuit, wax), life history, personal dispositions, encultured narratives from the surfing subculture and media, and so on' (Ford and Brown 2006: 162). However this assemblage conspicuously omits the place of surfing action, thus for a surf*er* to be surf*ing* the physical constellation of water, swell, weather system, continental shelf, sand bar, and reef are required to be part of this assemblage too. This coming together that constitutes the surfed wave as assemblage is therefore not simply a 'phenomenon but [also] a relationship between phenomenon' (after Virilio, cited in Rinehart and Sydnor 2003: 11); it is a constellation of 'things' into what Thrift (2004) might call a 'transient structure'.

Theorizing from the land in this way we can see how the place of the surfed wave can be understood, from the perspective of the surfer at least, as a place of assemblage. As with any assemblage, it is how each component part relates, connects and interacts that forms particular territorialized or watery places. In one proximate location on the sea a number of different places may occur one after the other: it may first be a place of flat calm, then a place of a wave, a place of a surfed barrel, a surfed wipe out, or a ducked-under wave. Each assemblage will produce a different relational agency, risk, and experience, before the constituent parts disengage and dismantle. This perspective emphasizes the changing nature of place in one location – due to the different combination of components in every instance, place is always dis- and re-assembling.

The Surfed Wave as Convergence

> There is a tremendous complicity between the body and the environment and the two interpenetrate each other. (Shields 1991: 14)

Whilst some surfers suggest a connection between mind, body, and sea in the assemblage of the surfed wave, others express their stoke in very different terms. Although the surfed place remains a coming together of the surfer (themselves an assemblage of board, wetsuit, wax, life history, etc) and the non-human, physical world (including water, tide, weather system, fetch, continental shelf, reef etc) many surfers describe their experience not in terms of *connections*, but in terms of *convergence*. Here, as Shields suggests above, surfers and waves are not simply connecting, but interpenetrating each other. This is described by surfers in the following ways:

> ... the ideal of *merging with the medium* ... of a now-expiring-and-never-to-exist-on-this-planet-again miracle. (Duane 1996: 66 emphasis added)

> I love the feeling of being in the sea and of riding a wave. I love the sea and
> so being able to spend time in it, and *be one with sea* is fantastic. (Interview
> Respondent, their emphasis)

These surfers express their involvement with the place of the surfed wave in
terms of being 'at one' with the amalgam of sea and swell, of 'merging' with this
'medium'. These affects do not refer to the execution of skills, or displaying the
intense concentration that is associated with 'flow' experiences, rather they refer
to a sense of union with the component parts of the surfed wave, a sense of losing
a coherent sense of self in part of something larger. This is possible due to the
radically different materiality of the sea when compared to the land (Steinberg
1999). This fluid materiality enables immersion within it (see Ingold 2008),
and facilitates a new sense of place emerging as a consequence. The following
respondent expresses this sense in this way:

> A sense of being a part of something that is timeless and much, much bigger
> than yourself, waves have been breaking since there has been water on the
> planet and that knowledge can ground me in a period of unease. (Questionnaire
> Respondent)

For these surfers being part of the surfed wave is not adequately described by
being one part of a larger assembly of components, but rather being subsumed by
a larger entity. As Scheibel notes, when joining a surfed wave, 'there appears to
be a disorientation in time and space where the surfer temporarily loses perception
of all external boundaries. There is an intensive and emotional reaction felt by
the rush of adrenalin to the muscles, with the resultant feeling of emotional
catharsis and the joyful sensation of having been so close in union with the ocean'
(1995: 256). The affective intensities produced through involvement with the place
of the surfed wave indicate a new way of framing these places; not as assemblages
but as convergences. Those involved identify relational interdependence and co-
constitution; their experience is not expressed in terms of 'this is how I feel in the
face of...' (after Game 1997), but rather 'this is how I feel being co-constituted
by ...', or, 'this is how I feel being converged with ...'. This co-constitution is
therefore the key difference between the surfed wave as assemblage on one hand
or convergence on the other. As Anderson and McFarlane reiterate, 'assemblages
are *not* organic wholes', their different parts are not 'subsumed into a higher entity'
(2001: 125, emphasis added). For these surfers, however, although the components
of the surfed wave may not have actually merged in their ontological form, there
is a sense to the surfer that they have done exactly that. The different parts of
the surfed wave have subsumed into a higher entity: there is no longer a surfer
and a wave, but a surfed-wave – for a short space of time this is now a singular

entity/process, ontologically joined from the perspective of the actor involved.[3] This framing therefore cannot be identified or experienced from the outside. To an onlooker, photographer, or academic, these places would seem like an 'ordinary' assemblage. However, for those who are a direct part of them, this coming together *feels* different, and unified.

The Implications of Assemblage and Convergence

This chapter has taken the opportunity provided by the relational turn to explore one particular human interaction with this watery world: surfing. By focusing on the surfed wave as an 'actor-centred geographic becoming', it has shown that the stoke experienced by the coming together of the surfer and the wave can be usefully understood using the notion of assemblage but also through adopting the notion of convergence. Assemblage frames both territorialized and watery places as a provisional assembly of components, joined together like parts in an engine or bricks in a child's toy-world. The notion of assemblage thus encourages a new way of looking at traditional places, whilst also encouraging the framing of new coming-togethers as places (for example, the surfed wave). In contrast, the notion of convergence suggests that when the component parts of the surfed wave come together they do not straightforwardly connect, rather their ontological form blends and blurs from the perspective of the participant. As Thomas (2007: 25) outlines, 'convergence fuse[s] thresholds'. From this actor-centred approach, the thresholds between subject and object, surfer and wave, are fused to make a coalesced, unitary entity/process. Although this convergence is transient, the emotion formed through it lingers, and becomes part of the world view and aspiration of those who experienced it. These experiences and the wish for further transient convergence become the provisional 'anchor' around which surfers orient their lives.

In the case of the surfed wave the difference between the connection of the assemblage and the blending of the convergence is significant. Once experienced, the blending of convergence affects world view and aspiration. Surfers who have experienced convergence view the world as ontologically unstable, with the possibility of merger both realistic and possible, as Californian surf-writer Weisbecker states: 'surfing forge[s] our perception of ourselves and of our relationship to the world around us' (2001: 11), with many desiring 'ideally to merge with the medium, this [is] our 'way of being in the world'' (Duane 1996: xiv). This aspiration to merge and emerge as part of a surfed wave, however temporarily, is known by 'Fuz', a South Walian surfer, as 'joined-up surfing'. He states, 'metaphors determine how we see the world, they influence our kids and

---

3   The term convergence therefore resonates here with the idea of *agencement* (Phillips 2006) from which the notion of assemblage was originally worked. Convergence re-emphasises the crucial aspect of blending and subsumption that has been lost in the dominant re-workings of the notion of assemblage.

life in general. You can become oblivious to your environment if you uncritically accept a language system … You become dull to the world. The point is to become aestheticized, to use your senses, to notice things' (Wade, 2007: 58). Through paying attention to the experience evoked through convergence, a new type of relational place is experienced through becoming part of the surfed wave. Coupled to modern metaphors, historical technologies, and scientific debates, surfers' perspective of the sea can therefore supplement and enhance our attempts to chart new maps of water worlds. Gaining insight from individuals' experiences of the place of the surfed wave also emphasises the importance of corporeal engagement with the sea. This embodied, practical engagement with the water world, in whatever form, is the key focus for Part II of this book: Ocean Experiences.

References

Anderson, B. McFarlane, C. 2011. Assemblage and geography. *Area*, 43(2), 124–27.

Anderson, J. 2009. Transient convergence and relational sensibility: Beyond the modern constitution of nature. *Emotion, Space, & Society*, 2, 120–127

Asakawa, K. 2009. Flow experience, culture, and ewll-being: How do autotelic Japanese college students feel, behave, and think in their daily lives? *Journal of Happiness Studies*, 5, 123–54.

Braun, B. 2004. Nature and culture: On the career of a false problem. In *A Companion to Cultural Geography*, edited by J. Duncan, N. Johnson, R. Schein. Malden: Blackwell, 151–79

Callon, M. 1986. Some elements in a sociology of translation. In *Power, Action, Belief*, edited by J. Law. London: Routledge, 19–34

Cresswell, T. 2004. *Place: A Short Introduction*. Oxford: Blackwell.

Csíkszentmihályi, M. 1990. *Flow: The Psychology of Optimal Experience*. Harper and Row: New York.

Delanda, M. 2006. *A New Philosophy of Society: Assemblage Theory and Social Complexity*. London: Continuum.

Deleuze, G. 1985. Nomad thought. In *The New Nietzsche*, edited by D. Alison. Cambridge: MIT Press. 143–9

Deleuze, G. 1993. *The Fold*. Minneapolis: University of Minnesota Press.

Deleuze, G. Guattari, F. 1981. *A Thousand Plateaux: Capitalism & Schizophrenia*. Minneapolis: University of Minnesota Press.

Doel, M. 1999. *Poststructuralist Geographies: The Diabolical Art of Spatial Science*. Lanham, MD: Rowman & Littlefield.

Doel, M. 2000. Un-glunking geography: spatial sciences after Dr Seuss and Gilles Deleuze. In *Thinking Space*, edited by M. Crang and N. Thrift, N. London: Routledge, 117–35.

Doherty, S. 2007. *The Pilgrimage*. London: Viking/Penguin.

Dovey, K. 2010. *Becoming Places: Urbanism/Architecture/Identity/Power*. Abingdon: Routledge.

Duane, D. 1996. *Caught Inside: A Surfer's Year on the Californian Coast.* New York: North Point Press.

Evers, C. 2006. How to surf. *Journal of Sport & Social Issues*, 30(3), 229–43.

Evers, C. 2009. 'The Point': Surfing, geography and a sensual life of men and masculinity on the Gold Coast, Australia. *Social & Cultural Geography*, 10(8), 893–908.

Ford, N. and Brown, D. 2006. *Surfing and Social Theory.* London: Routledge.

Game, A. 1997. Sociology's emotions. *CRSA/RSCA*, 34(4), 385–99.

Hayles, K., 1999. *How We Became Post-Human.* Chicago: University of Chicago Press.

Heidegger, M. 1956. *Being and Time* Oxford: Blackwell.

Hill, L. and Abbott. J.A. 2009. Surfacing tension: Toward a political rcological critique of surfing representations. *Geography Compass*, 3(1), 275–96.

Ingold T, 2008. Bindings against boundaries: entanglements of life in an open world *Environment and Planning A*, 40(8), 1796–810.

Jones, M. 2009. Phase space: Geography, relational thinking, and beyond. *Progress in Human Geography*, 33(4), 1–20.

Kampion, D. 1997. *Stoked! A History of Surf Culture.* Santa Monica: General Publishing Group.

Kampion, D. 2004. *The Lost Coast.* Salt Lake City: Gibbs Smith Publisher.

Latour, B. 1993. *We Have Never Been Modern.* Cambridge, MA: Harvard University Press.

Latour, B. 1999. P*andora's Hope: Essays on the Reality of Science Studies.* Cambridge, MA: Harvard University Press.

Laurier, E. and Philo, C. 1999. X-morphising: Review essay of Bruno Latour's Aramis, or the love of technology. *Environment and Planning A*, 31, 104771.

Malkki, L. 1992. National geographic: The rooting of peoples and the territorialisation of national identity among scholars and refugees. *Cultural Anthropology*, 7(1), 24–44.

Marcus, G. and Saka, E. 2006. Assemblage. *Theory, Culture & Society*, 23(2–3), 101–6.

Murdoch, J. 2006. *Post-Structuralist Geography: A Guide to Relational Space.* London: Sage.

Nazer, D. 2004. The tragicomedy of the surfers' commons. *Deakin Law Review*, 9(2), 655–713.

Oakes, T. 1997. Place and the paradox of modernity. *Annals of the Association of American Geographers*, 87(3), 509–31.

Phillips, J. 2006. Agencement/assemblage. *Theory, Culture & Society*, 23(2–3) 108–9.

Raban, J. 1999. Passage to Juneau. *A Sea and its Meanings.* London: Picador.

Rinehart, R. and Sydnor, S. 2003 Proem. In *To the Extreme: Alternative Sports, Inside and Out*, edited by R. Rinehart and S. Sydnor. Albany: State University of New York Press, 1–17.

Rose, M. 2002. Landscape and labyrinths. *Geoforum*, 33, 455–67.

Rouse, J. 1996. *Engaging Science: How to Understand its Practices Philosophically.* Ithaca: Cornell University Press.

Scheibel, D. 1995. Making waves with burke: Surf Nazi culture and the rhetoric of localism. *Western Journal of Communication*, 59(4), 253–69.

Sheller, M. and Urry, J. 2004. Places to play, places in play. In *Tourism Mobilities: Places to Play, Places in Play*, edited by M. Sheller and J. Urry. London: Routledge, 1–10.

Shields, R. 1991. *Places on the Margin.* London: Routledge.

Shields, R. 2004. Surfing: Global space or dwelling the waves? In *Tourism Mobilities: Places to Play, Places in Play*, edited by M. Sheller and J. Urry. London: Routledge, 44–51.

Steinberg, P.E. 1999. Navigating to multiple horizons: Towards a geography of ocean space. *Professional Geographer*, 51(3) 366–75.

Stranger M. 2011. *This Surfing Life: Surface Substructure and the Commodification of the Sublime.* Farnham: Ashgate.

Stranger, M. 1999. The aesthetics of risk: A study of surfing. *International Review for the Sociology of Sport*, 34(3), 265–76.

SurfingBlog, 2009. Trestles Waves. Available at: http://www.thesurfingblog.com/wp-content/uploads//2009/08/trestles-surf-sign.jpg [accessed, May, 2010].

Thomas, S. 2007. Littoral space(s): Liquid edges of poetic possibility. *Journal of the Canadian Association of Curriculum Studies*, 5(1), 21–9.

Thrift, N. 2004. Intensities of feeling: Towards a spatial politics of affect. *Geografiska Annaler*, 86(1), 57–78.

Vitterso, J., Vorkinn, M., and Vistad, O., 2001. Congruence between recreational mode and actual behaviour – a prerequisite for optimal experiences? *Journal of Leisure Research*, 3(2), 137–59.

Wade, A. 2007. *Surf Nation.* London: Simon & Schuster.

Weisbecker, A. 2001. *The Search for Captain Zero.* Los Angeles: Tarcher.

Wood, N., and Smith, S. 2004. Instrumental routes to emotional geographies. *Social and Cultural Geography*, 5(4), 533–48.

PART II
Ocean Experiences:
Embodied Performances,
Practices and Emotions

Chapter 6

The Day We Drove on the Ocean (and Lived to Tell the Tale About It): Of Deltas, Ice Roads, Waterscapes and Other Meshworks

Phillip Vannini and Jonathan Taggart

'Our world is a watery world', write Jon Anderson and Kimberley Peters in their introduction to this book, but despite this fact, geography has largely evolved as a landlocked discipline (also see Lambert et al. 2006, Peters 2010). Despite some positive developments in the fields of island studies (for example see Baldacchino 2007), the geography of the sea (Peters 2010, Steinberg 1999, 2001), and the social and cultural geographies of water (see Bear and Bull 2011 for a review) I cannot disagree with their view. In the past I[1] myself have attempted to transcend the terrestrial limitations of social and cultural geographies by focusing on ocean travel by way of ferry boats (Vannini 2012). However, my attention to the seas as liquid material places has been limited. When I was invited to contribute a piece drawn from my ferry research to this collection I felt the need to further develop my experience of a water world in a way I had not done before. However, my first attempts were frustrating and led nowhere. I felt I lacked the imagination and conceptual creativity to invert the land bias. Then, during a wayfaring stint I stumbled upon something promising.

While most of this book discusses oceans, I thought that if we are serious about reversing geography's land-centric bias we should not erect boundaries amongst diverse water worlds – such as lakes, rivers, and seas – or building further boundaries between land and water (also see Anderson, in this collection). Indeed if it is true that water, like land, is a space that shapes and is shaped by hybrid socio-cultural-physical processes (see Gandy 2004, Kaika 2004, Steinberg 1999, Swyngedouw 2009), those spatializing processes need to be examined in their precise physical and material details, and ideally in their lived experiences and felt practices. If water is not so unlike land, I reckon, then you should drive on it, and not just philosophize about it. As Jonathan and I were conducting fieldwork on off-grid living in Canada's Northwest Territories during the month of February 2012, the occasion to do so

---

1   Here and elsewhere in this chapter 'I' refers to the first author.

presented itself. This writing of a distinct embodied ocean experience is the result of participant observation and 15 interviews conducted in the Mackenzie Delta region of Canada's Northwest Territories, and Inuvik in particular.

'Canada's most easily accessible remote community' is not Inuvik's motto, but it might as well be. Situated two degrees above the Arctic Circle near the border between the Yukon and Northwest Territories, Inuvik is the unofficial administrative, educational, and economic capital of Canada's Western Arctic, and – as home of 4,000 people – one of the of the most populated towns in the world's circumpolar region. Reaching Inuvik by air is relatively costly (though inexpensive by Arctic standards), but as long as the weather cooperates it is entirely hassle-free. Flights to and from Inuvik depart almost daily, year-round, from Edmonton, Yellowknife, and Whitehorse.

If air routes do not suit your fancy, options to reach 'Canada's most easily accessible remote community' exist aplenty. For starters, you could try the famed Dempster Highway. Driving the Dempster will allow you to put to shame anyone who boasts of having 'survived' the Alaska Highway. From the Alaska Highway you raise the ante and double down onto the Klondike Highway, then approximately 40 km east of Dawson City, Yukon, you want to hang a right toward Inuvik: a mere 736 km north. And if this doesn't yet sound like a real adventure you could try boating down the Mackenzie River all the way from Great Slave Lake, or even further down, from Alberta's Athabasca River. Or for a different option you could hitch a ride from a Northwest Passage-bound supply ship leaving the Port of Vancouver and accessing the Beaufort Sea by way of Alaska.

Now, Inuvik is accessible by (unpaved) road from the rest of Canada, but its status as our country's northernmost community reachable year-round by road is not so clear. While one can drive the Dempster Highway to Inuvik in winter (weather permitting) or summer, it is impossible to do so during parts of spring and fall. During spring ice that forms bridges over the Peel and the Mackenzie River breaks up, but not in small enough chunks yet to allow navigation. Throughout the short fall season ice instead begins to take shape, impairing ferry navigation while not being consistent enough to allow driving. During these times, respectively referred to as 'break-up' and 'freeze-up,' Inuvik is only accessible via air routes. Other communities in the Mackenzie Delta – like Aklavik and Tuk – are also accessed by ice road only seasonally.

Driving the Ice Roads

'Ok, so what do I do?' I ask Jamie.

'Weeellll, that's the gas pedal…'

'Come on man, I'm serious; how do you drive on ice roads?'

'There's nothing to it,' Jamie reassures me with a smile, 'just take it easy on any curve; slow down to maybe 50. Keep your eyes on the road. And that's about it. You'll pick it up as you drive.'

'That's it?!'

'That's it.'

I am peeved. I feel cheated. What does he mean that's all there is to it? I have watched the first two seasons of The History Channel's *Ice Road Truckers* religiously. There allegedly are all kinds of dangers to steer clear of and all sorts of evil monsters to slay along the way. Like road-hogging semi-trailers who will bully you into snow banks and ditches just because you're on their turf. So I huff and puff, but I resign to the thought that I won't be able to claim any superhuman skill and I switch our rental 4x4 Dodge Durango in forward drive. No chains, no GPS. Just a mug of drip coffee and some expired, overpriced granola bars bought at Arctic Foods downtown. And we're off.

'Keep your speed between 60 and 80,' Jamie advises. That's fine with me. Though I see vehicles coming the other way easily clicking 100 or even 120 km/h we're in no hurry and our speed needs to accommodate Jonathan's camera work often taking place outside the passenger window. It's -20 degrees Celsius. We're here during a warm spell. The day's plan and itinerary are basic. Leave Inuvik at the crack of dawn, 10am. Drive on the Mackenzie River for 160 km, then keep driving onto the Beaufort Sea and keep aiming north for about 40 klicks. Exit at Tuk. Find Chucky; our local guide and gatekeeper. Ask him questions. Check out everything of interest. Then head back.

Jamie is our Inuvik-based guide and he's in charge of driving and saving our asses in case something goes awry. I'm fine with him taking care of the latter, but there is no way in frozen hell I ain't driving. If *Ice Road Truckers*' star Alex Debogorski can do it, I can too. Regardless of how easy it turns out to be. Oh, that wasn't a typo, by the way. I really did mean on and onto. To reach Tuk between mid-December and early May you don't drive by the Mackenzie River and by the Beaufort Sea. You drive on them.

Picture the Mackenzie Delta like this: fry some bacon, preferably of the fatty kind. Keep it on the pan for a while, then take the burned meat off and pour half a cup of water on the grease. The result will be an unpredictably complex, irregular pattern of round-ish blotches and worm-like canal-type shapes sinuously winding their way around the pan. Now, for a different chromatic visualization of the area, but one that looks astonishingly similar in shape, head for Google Maps on the Internet and look at the Mackenzie Delta map. You'll get pretty much the same pattern, only different colors.

If *Ice Road Truckers* lied about how perilous it is to drive on an ice road, Google Maps tells an even more ostentatious lie about the visual appearance of the region. For Google Maps blue means water, and grey means land. Water, in turn, means lakes, rivers, and seas, whereas grey means earthly dirt. Clear enough distinctions. But that's a partial representation of truth at best. During winter grey and blue are nothing but white here. White everywhere. We southerners are accustomed to a world shaped by the easy distinctions cooked up by our minds and reinforced by cartographers and other editors of our imagination. Blue or grey. Water or land. Rivers or lakes. Oceans or continents. With those simple distinctions in mind we act as if they were real at all times. But they are not real here, not now. The Delta is all white during winter: sky, surface, and contours. Water is everywhere in different states, solid, liquid, air, deconstructing boundaries, and forcing us southerners to break them apart too, and act accordingly. Indeed, flash forward to one day later on the road to Aklavik …

'Hey… Hey! Phillip, what are you doing?'

'Steer left, steer Left! – Were are you going?!'

'Oh, no! Damn it…!'

'I'm sorry guys, I'm really sorry. I lost sight of the road. It's all fucking white everywhere! Is everyone ok?'

'Yes, but shit! What do we do now?'

'Yeah what do we do now?' – I ask William, our guardian angel from the Aurora Research Institute, sent to protect us on our way to Aklavik.

'Don't worry, that's why we brought a rope and a satellite phone, right?' He replies, unperturbed, as if nothing happened. His call to his coworker back at the Research Institute, however, fails to reach him.

'Man, this is all my fault! It's so embarrassing. What do we do now?' I mumble in alarm.

Even less discomposed than before William waves at a driver slowing down to check on us. He tells them we're ok and sends him on his way. Our friend is coming to help us, he says. Even though we haven't even reached him yet. I suppose I would be more agitated if we had gotten stuck farther out of town. Thankfully we are no more than a five minute walk from the only gas station in Inuvik, and right at the major intersection where the paved road ends, turns onto a bank of the Mackenzie, and descends onto the ice road. There can hardly be a worse spot to get stuck one if you're the type of southern researcher who worries about being the town's laughing stock, through there can hardly be a better one if you're concerned about rescue. Yet, William's calm is unsettling to me.

'You're so calm, William,' I observe. 'Do people get stuck in snow all the time around here?'

'Yeah, it happens all the time,' he responds calmly, while slowly hitting re-dial on the phone and timing his punch-line, 'though we generally make it onto the road first.'

I blush. Jon erupts in laughter. Within seconds another large truck drives by and stops to check on us. It's Moses – William's younger brother, of all people. 'What's the matter, guys?' he asks, leaning outside the window of his muscular Ford F-Series pickup.

'I got us stuck' I reply, consternated and sheepish.

'I can tell,' Moses replies. He pauses for a moment, then unleashes his own wisecrack on me: 'No native eyes, eh?!'

Jon and William fall on their knees laughing. I briskly change the topic to ropes and towing. Within minutes we're out of the pile of fresh snow.

One common way of examining how systems of material objects coalesce together is by utilizing the concept of assemblage. An assemblage is essentially a constellation of different objects which are drawn together to comprise a whole (deLanda 2006, Robbins and Marks 2009). Just like a constellation of stars is a whole of objects that we understand to be in relation to one another, constellations of objects comprise articulated wholes that are more than the sum of their parts. Research on assemblages shows that their material qualities are of paramount importance (see Robbins and Marks 2009). Materials are not unimportant or easily replaceable. They are distinct, unique, and have a life of their own. Materials, writes Ingold (2011: 28), 'are the active constituents of a world-in-formation.' 'Wherever life is going on,' he continues, 'they are relentlessly on the move – flowing, scraping, mixing, and mutating.' As materials move as part of their life processes they intersect with one another and often undergo transformations as a result. Think of what happens, for example, when bodies of water meet currents of cold air.

In turn, as humans go about their daily businesses their lines of movement intersect with various materials. All these intersecting threads of activity end up over time forming meshworks of 'entangled lines of life, growth, and movement' (Ingold 2011: 63) which constitute the places we inhabit. Meshworks are a type of assemblage, a type distinct from networks. At times it makes sense to think of meshworks as metaphors for interaction between the human and the non-human world. But at other times meshworks are more material than metaphorical – existing and working as actual knots issued forth from a world in constant formation, a becoming world in which material flows and human movements intersect in a constant unfolding of events (see Cresswell 2002).

Ice roads are a meshwork: an intricate assemblage of trails carved by human movement and by the various movements of water across the seasons. They are not an inert container of movement taking place over them by rolling vehicles like trucks and cars. Such a Pointilist view (see Doel 1999) would miss the relatedness of ice roads and the bodies of water they are channeled by. It would miss the embeddedness of ice roads within their environment. The ice road meshwork is a mutating, temporary, ephemeral, landscape (cf. Dewsbury 2000: 487). Its existence is an effect of 'differential spacings' (Doel 1999: 12) emerging from the changing relations between ocean, lakes, rivers, and the air. Indeed its formation is an event; an event whose experience problematizes the very nature of landscape – turning water into a land-like *icescape* and into constantly changing relations.

As river and ocean water meets cold air ice roads form and change. As sun rays shine on ice roads their surfaces change. As water accumulated in the shape of snow meets winds, snowdrifts begin to form, confusing distinctions between road and snow banks. And so forth. So if congealment is an event leading to the territorialization of the ice road meshwork, then liquefaction is an event leading to its deterritorialization and to the emergence of a delta that is no longer drivable but navigable. Whether congealed or liquefied, frozen solid, muddy, or in the midst of breaking up or thawing out, the water-land-air meshwork reveals to us a domain of entanglement open to different relations with humans for different access assemblages. Such a tangle reveals to us 'the texture of the world' and in bodily inhabiting this tangle of constantly changing, interconnecting trails beings 'contribute to its ever-evolving weave' growing 'along the lines of their relationships' (Ingold 2011: 71).

The road to Tuk is astonishingly ordinary. It is ordinary because, in all honesty, it is nothing but a road. As wide as a three-lane highway, with no stop lights or stop signs, trafficked by no more than one vehicle every ten minutes or so, and policed by no one but moose and low-flying ravens (though speed traps are not unheard of), the Mackenzie River ranks as one of the most comfortable and spacious routes I have ever driven. On the other hand, it feels astonishing to drive over any kind of substance that a map paints blue, and this sense of astonishment must be accompanied by a re-skilling of my vision (cf. Grasseni 2004) if I want to get around safely. There are a few reasons why I feel this way.

For starters the consistency of the road is entirely unusual to me, and while not frightening, it is unsettling and unfamiliar. Asphalt-based pavement is generally smooth, uniform, and during hot days even sticky. Ice pavement is the complete opposite; it is impenetrably hard, unevenly striated, and surprisingly dry – the colder it is, the drier it is. To be sure, imminent dangers of ice cracking as we drive by – fueled by *Ice Road Truckers*' dramatic graphic design tricks – are no longer harassing my imagination. I know that every year the Northwest Territories Transportation Department carefully selects opening and closure dates of each ice road to ensure drivers' safety. Average dates are punctually reported on the web. December and May are the shoulder months. That's when you need to be extra careful as the ice hasn't thickened enough, and overflow causes all kinds of problems. But during February, the time we are here, trucks of up to 30,000 kilos and gauche West coasters driving rental SUVs can travel worry-free. So it's just a matter of forming new habits.

Secondly, there is an unsettling feel accompanying the movement on – or perhaps I should really say *with* – the ice road. In my mind paved highways are relatively unresponsive to my driving. I know that repeated driving over long periods of time can cause ruts on paved surfaces, but for the most part the material of which the pavement is comprised is of no immediate concern to a driver as it feels stable, immobile, and unchanging. Ice roads are radically different. With every vehicle driving by, the ice surface changes in consistency a bit. Most notably, tires sweep snow off the ice surface. Snow works as a warming blanket over ice, raising temperatures underneath. Thus driving works as a road cleanser – making the ice colder and drier. There is also an issue of weight. As the driver of a light SUV I am only mildly concerned about this, but I know that drivers of 18-wheelers need to take into account sloshing much more seriously. There is water underneath the ice surface. As you drive you create a wave. If you drive a heavy vehicle at high speeds, that wave can become sizeable enough and fast enough to catch up with you, which can cause overflow and even cracking, especially near turns and river banks. And tides are also an issue. When driving on the sea, changing tides can cause the water underneath the ice to rise or fall, thus provoking cracks – or more commonly sinkholes and overflow – on the surface. Besides all this it is quite easy for patches of drift snow to accumulate on the road and create bumps.

And thirdly, you need – you really, attentively, uncomfortably need – to keep your eyes on the road. Driving in normal conditions down south for me is a bit like driving on automatic pilot. This tunnel vision mode is a way of avoiding visual fatigue. It relies for its functioning on the sharp chromatic differences I have become accustomed to while driving. But here the experience of the situation is different. On ice roads there are worse driving moments and there are better ones. When the sky is clear or partly sunny it is easier to visually separate surface from atmosphere, foreground from background. This allows me, as a driver, to keep my eyes just on the ice road, constantly alternating between the tiny strip of surface my wheels are about to roll over and the snow banks on the sides. A blue sky quickly recedes into the background, unworthy of my attention. In addition, its light casts sharp shadows – making snow banks stand out from the road in sharper relief and thus allowing me to see curves far in advance. If the road has been recently grooved, and if no snow has drifted on it, my visual task becomes even easier because the ice surface is characterized by a beautiful hue of grey-blue.

However, the road strip is not always so wide and not always so blue. During moments of bad visibility, the sky, snow banks, and drivable surface blur to almost indistinguishable hues of white, making driving feel like a Gestalt switch that is too dangerous to switch back from. Even on cloudy days there are times when the Arctic sky is still filled with much brilliant light. Whereas in coastal British Columbia a cloudy day generally means a dark-grey overcast sky, the cloudy Arctic sky feels undercast, seemingly blending its light with the land, drifting snow, and frozen water, absorbing into one another like flour mixes with egg whites and sugar. In the open tundra and the open sea, where there are no trees or hills in the distance, limited visibility is especially crippling. Driving the last 40 km on the Beaufort Sea on a foggy day on the way to Tuk feels like rolling on an absolutely flat, white billiards table with nothing but solid, number-less, white balls around you. All you can do is visually hang on to the last few remaining traces of the grooves semi-covered in snow and hope that any second now you will reach the top of the planet, get some kind of panoramic perspective from there, and start heading down the curvature of the earth from that point on.

Wayfaring

Human beings are the producers of their own lives. Yet, they are neither the only inhabitants of our world in possession of this productive power, nor are they omnipotent at imposing their own designs on our planet. Indeed it might be better to say that the human inhabitants of this world can only work with the materials they have available and – if they are wise – can and should only work with the ways the materials themselves work. Rather than imposing a design of their own on the environment, it makes sense to view our species as playing a part 'in the world's transformation of itself' (Ingold 2011: 6).

If water is what you have available to make a road with, then you might as well learn to work with the way water works. And as it turns out water works in many different ways, under many different conditions. As Strang (2005: 98) observes, the key material feature of water is that it is 'endlessly transmutable, moving readily from one shape to another.' Water is 'unruly' (Jones and MacDonald 2006, also see Budds 2009, Linton 2008). It is very resistant to deformation and disintegration, as opposed to rock, for example. This makes it a very challenging material to build roads with. The construction of conventional paved roads requires we combine rock-based aggregate particles (like stone, sand, and gravel) and mix them with asphalt. Asphalt – a petroleum derivate – glues together the different materials and reduces the heterogeneity of the different materials involved. Cold air works in a similar way with water, increasing its viscosity without reducing its potential for flow (Deleuze and Guattari 1986).

The construction of roads with water depends on water changing its state from liquid to frozen solid, and therefore on the particular relation between water and cold air that we call congealment. Congealment obviously depends on precise atmospheric conditions: both air and water must be cold enough for a sufficiently long period of time to allow people to use ice as a driving surface. In the Mackenzie Delta region ice is not stable year-round. Winter congealment there gives way to gradual spring thawing. By summer time the ice is fully liquidified and water can be used for access by way of boat navigation. So, water is obviously a material we can do things with, but it is also a constant process. What this process reveals to us is that materials are not homogenous or stable, but rather in flux. Water is constantly passing through stages of precipitation, evaporation, sublimation, congealment, precipitation, and liquefaction – thus being fully on the move. As Strang eloquently writes:

> water is always undergoing change, movement and progress. Captured in a cup
> or pond or lake, it evaporates or escapes and runs away: it is always physically
> flowing from one place to another in streams, torrents, waves and currents. Even
> in the calmest of conditions, its qualities are such that it reflects the most subtle
> changes in light, and so shimmers with movement. (2005: 98–9)

A map accounts for these processes very poorly. Google Maps and Google Earth, as we have seen, portray different bodies of water with same colour, regardless of the differences amongst them. They also portray water as constantly blue, regardless of its chromatic change across seasons. Google Maps and Google Earth in fact reflect a common attitude that most people seem to hold toward water, an attitude that discounts its diverse forms (preferring instead to focus on how it differs from land) and that neglects its processual qualities. Even I had this attitude when I first arrived in the Mackenzie Delta: water (in the form of ice) was something you weren't somehow meant to drive on, I felt. It was something that was clearly distinct from land; an outlook that made my driving on ice roads very arduous at first.

But as it turns out water – like the world as a whole – is constantly undergoing transformation, continuously generating diversity. It is by recognizing the transformational flows of water and by working with the uniqueness of these changing materials that inhabitants of the Mackenzie Delta region (and us visitors) can use roads and thus access each other's communities and the rest of Canada. It is by taking part in the region's constant self transformation that they create and follow their roads, routes, and trails. This brings me to an important junction in the line of my argument, one that I will further elaborate upon in a bit. But for now I want to simply prepare for that conclusion by outlining what we have learned so far.

In concrete, empirical terms we have learned that water, land, air are not easily distinct in the Mackenzie Delta. Ice roads epitomize the strength of the relations amongst water, land, and air as they exist as an assemblage of these three materials. This assemblage is not something that outsiders like me are readily equipped to handle. Map-makers, for example, fail to recognize how tightly knotted together these materials are. Their cartographic representations of the place – with their neatly demarcated chromatic boundaries – are evidence of this. And drivers from the south are just as naive. For some – like the producers of *Ice Road Truckers*, for example – ice roads are dangerous because water and ice are allegedly unlike land, and therefore foreign, unreliable, and treacherous (Steinberg 1999). For others, like me, ice roads are just uncomfortable because they are not characterized by the same distinctions (e.g. like chromatic ones, or those between land and water surfaces, or between foreground and background) and the kinds of stability they are used to. But what we have also learned is that despite some initial discomfort and inconvenience (like my eyes getting tired, or my driving into the ditch), adaptation is possible once a 'switch' is made. It is this switch that I want to invite geographers to make and to which this book more broadly invites geographers to make.

Google Maps practices another sleight of hand on us by failing to account for the Delta's ongoing transformation of itself. To get a vivid understanding of what I'm saying, go back to your earlier search and switch this time from the map view to the earth or satellite view. What is the weather like? What is the season? But is it always a clear summer day around here? A delta is a remarkable example of a meshwork, as the entanglements it is made of are lines undergoing constant movement and continuous change. Deltas, anywhere, are utter messes where it is nearly impossible to make the usual distinctions that our minds have become accustomed to (such as between water and land).

Ice roads are like deltas and tidal areas (Jones 2011): something to be imagined not in terms of fixity and separation but rather in terms of their fluidity and associations, spatial formations coalescing as 'temporary placements of ever moving material and immanent geographies … without prescribed or proscribed boundaries,' to hijack an idea from Amin (2004: 34). They mess with our embodied experiences because they are live – constantly changing in a way that we are not readily equipped to see objects behave. And they mess with our minds and bodies because while driving on them we easily forget we are moving with them. When you drive on an ice road for the first few times the difficulty in adjusting to the unique driving conditions comes precisely from the confusion arising from the loss of usefulness of our typical categories. That is why I drove us into a pile of fresh snow, and that is why I found it so difficult to tell apart foreground from background. It all felt a bit like shooting at a moving target while blindfolded.

Driving ice roads has always fascinated me because water – in my rather conventional view of the world – has always been clearly demarcated from land. Driving on water and especially on ocean water feels like an aberration of the natural order. That's because land, roads, and soil seem solid, they feel stable, and in contrast lakes, rivers, and seas are characterized by an uneven surface which does not have the same properties of firmness, density, and stability. But this common attitude toward the water world is undoubtedly 'terrestrocentric' (see Dalby 2007), and typical of us southerners and westerners. We view waterscapes for the way they differ from landscapes not for the ways they are similar. What would happen if we instead viewed the land from the perspective of water? What would happen if we wanted to see similarities and overlaps between land and water, rather than distinctions and boundaries?

Take the Dempster Highway for example. A normal map draws it as a distinct, firm line that has been laid on the land. But from our new viewpoint it makes sense to see it as winding its way through both land and water, in constant movement across the seasons – not unlike water flows. Indeed during winter reaching Tuk from the Yukon is only made possible by the way water and soil flow and bind together – ceasing to exist as separate entities, at least from the practical perspective of access. As snow packs and ice forms, the water in the land (as ice) and the water in the atmosphere (as snow) blend together. This allows for trails to be carved, for ice roads to be woven, and for driving to be possible. During the other seasons as well it makes sense to see water and land as a mesh, as a grid of layers of rock and pools of water. The surface of the Dempster is so hard on tires because it had to be built on crushed rock rather than gravel. The active layer of permafrost sitting underneath the crushed rock melts during the warm season, and that would cause gravel to sink. But even as crushed rock the road is still not impermeable from water. Water is everywhere: from standing pools and adjacent ponds and lakes (which at times flood the road), to puddles of mud, and to the Peel and Mackenzie Rivers which the Dempster flows into. The Delta and its ongoing seasonal transformation allow us to comprehend how deeply interconnected water, land, air, and people are.

So, what are we to make of all this? What can we do to make that switch simple and ignite our imagination? Drawing from the work of Tim Ingold again, I want to introduce a metaphor, that of wayfaring. Understanding how wayfaring works can allow us to make the 'switch' I talked about and look at places from a more fluid, water-like vantage point. Wayfaring refers to an unsettled and relatively unplanned way of journeying. Its antonym is that of transporting. Transport is planned in advance and allows no deviation from a settled route from point A to point B. Wayfaring is instead improvised and open to experimentation and extemporaneous creativity. Wayfaring is therefore not just an experience and practice of movement for the sake of movement, but also an experience and practice of encountering, discovering, learning, and knowing. Ingold (2007, 2011) writes that people learn by going around, by encountering lessons along paths of discovery, by weaving together tidbits of knowledge through stories in a continuous process of exploration.

To live in our world means 'to be embarked upon a movement along a way of life' (Ingold 2011: 12). As we follow and blaze trails, like a wayfarer uncovering new directions and abandoning old routes along the way, we develop skills and accumulate lessons, revise obsolete understandings and make new observations. Knowledge is therefore a path of navigation, and knowing is therefore a way of moving. By surfing the globe, however, what we uncover is not a rigid and unresponsive environment: 'the forms and spatial relations of the world around us are clearly not static and fixed; they are constantly being altered, updated, and constructed in ways that alter sociospatial relations' (Dodge and Kitchin 2005: 172). Any boater knows that even the same strait changes as different conditions (e.g. wind, tides, etc.) change. Wayfaring is therefore not a way of discovering inert environments 'that lie there in place without trajectories' (Massey 2004: 227) but a way of moving along with the ways that places themselves change and move. Ingold provides us with an especially poetic passage to grasp all this, when he writes that 'to be sentient' is to:

> open up to a world, to yield to its embrace, and to resonate in one's inner being to its illuminations and reverberations. Bathed in light, submerged in sound and rapt in feeling, the sentient body, at once both perceiver and producer, traces the paths of the world's becoming in the very course of contributing to its ongoing renewal. (2011: 12)

Geography's land-centric bias has, at times, caused us to lose our feel for the fluidity of spatial relations, especially water-based ones. This has prompted several contemporary perspicacious commentators to lament that the world appears to be 'understood in fixed and bounded ways ... in a way that effaces the vectors, the speeds and the differences between movements' (Adey 2006: 78). While it's easy to be a wayfarer at heart, it's easier to settle down and seek the comfort and convenience of old stomping grounds, familiar paths, and established categories. Humans are habitual creatures, and at times the very prospect of wayfaring feels laborious, tiring, and uncertain. Lightly worn paths then become well-marked ways of life. Canals become tunnels. Trails become roads. Roads become freeways with fewer and fewer exits. In that process the world and our individual existences become place-bound and full of clear boundaries, like those between the solid and the fluid, blue and white, oceans and continents, land and water, and so on (see Lewis and Wigen 1997, Steinberg 1999, Westerdahl 2005).

But those boundaries are more revealing of the lines we are willing to cross, and the lines we are used to crossing, than real lines demarcating different realities. During our day-to-day existence we tend to forget this fact. This forgetfulness of what Ingold (2011) calls the *logic of inversion*. The logic of inversion is a logic born out of laziness; an attitude that views the world as a simplified, and yet unnaturally complex entity. But in the process of getting back on the road, of wayfaring once again, the logic of inversion is inevitably confronted and challenged over and over. The world is more complex, and yet simpler, than the way we have become accustomed to imagining it and using it. Processes of territorialization and reterritorialization help us see these processes, but their viewpoint is still, often, that of *terra firma*, of the terrain, the territory, the land. A water world is more fluid than that, and its 'vibrant materiality' (Bennett 2010) in form of constant liquefaction, evaporation, congealment, sublimation, precipitation, and thawing is something we fully need to come to terms with.

We learn as we go. We find our ways along paths of exploration. We journey together with other inhabitants of our world: inorganic materials, plants, and animals. Through our practice and experience we learn that the way we generally represent life on this world is only partly adequate: full of shortcuts, lazy habits, faulty bits generated by an inverted logic unable to understand how solids can change (Thrift and Dewsbury 2000). By wayfaring once again – much like born again children, with the same curious disposition toward wonder, with the same thirst for astonishment – we return to a world that is continuously unfolding, brimming with life, dripping with complex but elegantly simple relations, subject to endless flow, transformation, and change. At first we reject such realization and force our way into and through it. We try to channel our ontological and epistemological designs, we fail to observe how materials work in their own way, how they move around us, with us, forming organic meshwork assemblages. But if we are courageous enough to persist and follow new trails, observant enough to learn from practice, from working with the meshworks of the world – rather than in spite of them, then we should be able to revert the logic of inversion. And that's the switch: the reversion that's needed if we truly wish to have not just 'blue theory' (Dalby 2007) but a white, blue-gray, and brown one too, and not just a geography of the oceans or of water, but oceanic, littoral, fluvial, glacial, and watery geographies: fluid meshwork-like entities that are fully on the move, open to unexpected discoveries and the occasional storm, like water itself.

Acknowledgement

We wish to thank the Social Sciences and Humanities Research Council of Canada for their funding and the Aurora Research Institute for their gatekeeping, support, and logistical assistance.

References

Adey, P. 2006. If mobility is everything then it is nothing: Towards a relational politics of (im)mobilities. *Mobilities*, 1, 75–94.

Amin, A. 2004. Regions unbound: towards a new politics of place. *Geografiska Annaler Series B*, 86, 33–44.

Baldacchino, G. (ed.). 2007. *A World of Islands*. Charlottetown: PEI: Acorn Press.

Bear, C. and Bull, J. 2011. Water matters: Agency, flow, and frictions. *Environment & Planning A*, 43, 2261–6.

Bennett, J. 2010. *Vibrant Matter: A Political Ecology of Things*. Durham, NC: Duke University Press.

Budds J. 2009. Contested H(2)O: Science, policy and politics in water resources management in Chile. *Geoforum*, 40, 418–30.

Cresswell, T. 2002. Introduction. In *Mobilizing Place, Placing Mobility: The Politics of Representation in a Globalized World*, edited by G. Verstraete and T. Cresswell. Amsterdam: Rodopi, 195–218.

Dalby S. 2007. Anthropocene geopolitics: globalisation, empire, environment and critique. *Geography Compass*, 1, 103–18.

deLanda, M. 2006. *A New Philosophy of Society: Assemblage Theory and Social Complexity*. London: Continuum.

Deleuze, G. and Guattari, F. 1986. *Nomadology: The War Machine*. New York: Semiotext(e).

Dewsbury, J.D. 2000. Performativity and the event: Enacting a philosophy of difference. *Environment and Planning D: Society & Space*, 18, 473–96.

Dodge, M. and Kitchin, R. 2005. Code and the transduction of space. *Annals of the Association of American Geographers*, 95, 162–80.

Doel, M.A. 1999. *Poststructuralist Geographies: The Diabolical Art of Spatial Science*. Edinburgh: Edinburgh University Press.

Gandy, M. 2004. Rethinking urban metabolism: Water, space and the modern city. *City*, 8, 363–79.

Grasseni, C. 2004. Skilled vision: An apprenticeship in breeding aesthetics. *Social Anthropology*, 12, 41–55.

Ingold, T. 2007. *Lines: A Brief History*. London: Routledge.

Ingold, T. 2011. *Being Alive*. London: Routledge.

Jones, O. 2011. Lunar-solar rhythmpatterns: Towards the material culture of tides. *Environment & Planning A*, 43, 2285–303.

Jones, P. and MacDonald, N. 2006. Making space for unruly water: Sustainable drainage systems and the disciplining of surface runoff. *Geoforum*, 38, 534–44.

Kaika, M. 2004. Interrogating the geographies of the familiar: Domesticating nature and constructing the autonomy of the modern home. *International Journal of Urban and Regional Research*, 28, 265–86.

Lambert, D., Martins, L., and Ogborn, M. 2006. Currents, visions, and voyages: Historical geographies of the sea. *Journal of Historical Geography*, 32, 479–93.

Lewis, M. and Wigen, K. 1997. *The Myth of Continents: A Critique of Metageography.* Berkeley: University of California Press.

Linton, J. 2008. Is the hydrologic cycle sustainable? A historical geographical critique of a modern concept. *Annals of the Association of American Geographers*, 98, 630–49.

Massey, D. 2004. Geographies of responsibility. *Geografiska Annaler Series B*, 86, 5–18.

Peters, K. 2010. Future promises for contemporary social and cultural geographies of the sea. *Geography Compass*, 4, 1260–72.

Robbins, P. and Marks, B. 2009. Assemblage geographies In *SAGE Handbook of Social Geographies*, edited by S. Smith, R. Pain, S. Marston and J.P. Jones, III. London: Sage, 176–93.

Steinberg, P. 1999. Navigating to multiple horizons: towards a geography of ocean space. *Professional Geographer*, 51, 366–75.

Steinberg, P. 2001. *The Social Construction of the Ocean.* Cambridge: Cambridge University Press.

Strang, V. 2005. Common senses: Water, sensory experience, and the generation of meaning. *Journal of Material Culture*, 10, 92–120.

Swyngedouw, E. 2009. The political economy and political ecology of the hydro-social cycle. *Journal of Contemporary Water Research and Education*, 142, 156–60.

Thrift, N., Dewsbury, J.D. 2000. Dead geographies – and how to make them live. *Environment and Planning D Society & Space*, 18, 411–32.

Vannini, P. 2012. *Ferry Tales: Mobility, Place, and Time on Canada's West Coast.* New York: Routledge.

Westerdahl, C. 2005. Seal on land, elk at sea: Notes on and applications of the ritual landscape at the seaboard. *The International Journal of Nautical Archaeology*, 34, 2–23.

Chapter 7

What I Talk About
When I Talk About Kayaking

Jon Anderson

Wherever there is a channel for water, there is a *road* for the canoe. (Thoreau 1864: vii14, my emphasis)

A week on the *road* is enough to confirm my instinct that th[ese channels], with [their] climatic and ecological shifts, [produce] a different consciousness. Our west is their east. … There are new orthodoxies …" (Sinclair 2012: no page, my emphasis)

Introduction

As Thoreau suggests, the practice of kayaking can be undertaken in many ways and on many bodies of water. The presence of any 'channel' of water is enough to let you float your kayak-boat. The (relatively) straightforward act of flotation and subsequent self-propelled movement across a body of water can occur on lakes, in rivers, or on the sea, as well as on specially constructed pools or water courses, and even on unplanned routes over flooded land. This act of buoyancy, balance, and propulsion is scientifically similar in all these various water worlds.

Beyond this similarity, however, kayaking is a many and varied pursuit. In general the activity is split into three main disciplines: kayaking, sea kayaking and open canoeing. Whilst canoeing involves a single paddle and an open seating area or cockpit, kayaking refers to 'those canoes where you're secured in with a waterproof seal and use a double-bladed paddle' (Cooper 2002). In terms of kayaking and sea kayaking (as opposed to canoeing), each of these activities is very different too. As Rowe states:

The cynic might say that modern kayaking has become a collection of totally unrelated sports – that the flat water sprinter has nothing whatsoever in common with the white water freestyle star, and the rambling sea kayaker shares nothing at all with the competitive slalom racer. (2002: 3)

Kayaking is therefore not an essentialist activity in its form, purpose, or location. It can be practiced for competitive sport on rivers or artificial courses, as a leisure pursuit on the sea or other water channel (see Hanson 2001), and remains a key means of transport and survival in many water-dependent indigenous societies (see Walls 2012). In many ways therefore, how kayaking is experienced and understood is an 'actor-centred activity' (Anderson 2012a, Jones 2009). In theory, there are as many ways to codify and frame the activity as there are paddlers. Even in terms of kayaking in coastal and sea waters there are a multitude of experiences available. As Goodman states:

> Sea kayaking is about taking journeys: trips around the bay on a sparkling summer's day; exposed passages along lines of cliffs with dark sea caves to explore; around promontories that spawn tide races where ten-knot currents can kick up huge waves on the calmest of days; gruelling crossings when settled weather and trained muscles are essential; crossings accompanied by dolphins and broaching whales; journeys in foul weather when the hiss of spray sets the adrenaline racing and the final landfall to safety through the surf is exhilarating and chilling to both mind and body. The physical list is endless. (Goodman, in Duff 1999: xi)

Despite these necessary caveats, this chapter seeks to outline some experiential understandings of the practice of kayaking on the coastal ocean. Rather than focusing on a particular watery world – a specific geographical *place in the sea* (perhaps one coastline, one current, or one tidal flow for example); it rather draws our attention to the *space of the sea* itself, and the experience of kayaking upon it. With apologies to Murakami (2009), this chapter outlines what kayakers 'talk about when they talk about sea kayaking'. It discusses what it feels like to be buoyant on the ocean, balanced only by a combination of kayak-hull and appropriately-positioned body weight, and propelled across the water surface solely by double-oared paddles. The chapter achieves this by drawing on the experiences of coastal and sea kayaking from three key sources: a survey conducted with 395 kayakers based in South Wales, UK,[1] the autobiographical writings of Chris Duff (who has paddled over 12,000 miles on the sea and completed circumnavigations of the UK, Ireland, and South Island New Zealand), as well as my own experiences of over 10 years coastal kayaking. From these sources, the chapter will make a number of points. Firstly, that sea kayaking can be understood as part of the

---

1 This survey was conducted with kayakers frequenting the Cardiff International White Water (CIWW) Centre in Cardiff Bay in 2011. Cardiff International White Water is an Olympic standard white water centre and the first on-demand white water rafting centre in the UK. An online survey was written by the author and CIWW staff member Bryony Rees, and was sent to 2,000 kayakers on the Centre's mailing list. The survey was completed by 395 people representing a range of ages, genders, skill levels, and experiences. Ninety-eight percent of all respondents had paddled on both artificial and natural bodies of water.

family of extreme sports; it is at once sociable, adrenalin-fuelled, and – due to the unique elemental qualities of the sea – full of risk. This search for thrill and risk in watery environments makes it an addictive pursuit, and as a consequence, sea kayaking becomes part of many participants' embodied identity. Secondly, the chapter argues that although kayaking may be similar to other extreme sports in many ways, there is something specific about the water that makes this practice different from land-based pursuits. Due to the influence of the water world, the chapter explores how sea kayaking can be considered not just as an extreme sport, but also a form of artistic performance. As a consequence, and following Sinclair (2012), kayaking on the ocean 'road' can generate a 'different consciousness' and a 'new orthodoxy' among participants. In short, I argue that sea kayaking can offer participants a new way of looking at the world; in the words of paddle-maker Werner Furrer, kayakers 'find themselves with a new paradigm' with which to understand their relations to both water and terrestrial worlds (in Duff 2003: iv).

The Sea as Viewed from the Land

I used to look out from my window and see the vast horizon of Cardigan Bay, Wales. On many occasions this view was filled with the ocean churning, with sets of waves breaking in angry and vicious multitudes. On other occasions, the same sea was stilled, the oceanic expanse as calm as a windless pond. Many people, from Melville's Ishmael (1992) to modern day surfing nomads (Brownley 2010), tell how they are 'called to the sea'; they are summoned to view the sea not solely from their window on the land, but to engage directly with its varying moods and rhythms. I am no different; however, unlike Ishmael or surfing nomads my preferred mode of engagement is through a kayak and a paddle. The call to kayak is heard by a growing number of individuals (see Cooper 2002) and the ease of the activity (see Hanson 2001: 9) has garnered a wide following.[2] Despite the simplicity with which a newcomer to the sport can pick up the basics in calm conditions, sea kayaking has many of the characteristics of an extreme sport. Extreme or 'whizz sports' (Midol 1993, Wheaton 2004) include activities such as rock climbing, sky-diving, skateboarding, surfing, and kite-surfing, and all share a degree of 'family resemblance' in the Wittgensteinian sense of the term (Tomlinson et al. 2005, van Bottenburg and Salome 2010). Extreme sports are individualistic in nature (as opposed to team-oriented); they are non-aggressive;

---

2   As an illustrative exemplar, according to respondents of the survey at CIWW, kayaking appeals to a wide age group. The youngest kayaker who responded to the survey was 9 years old, the eldest 73, with the average age of respondents being 32 years. Fifty-six percent of all respondents had been active in the sport for 10 years or more (with 28% participating for over 20 years). Eight percent of all respondents classed themselves as 'beginners', 52% as 'intermediates', and 40% as 'advanced' kayakers.

participatory rather than spectator-focused; consuming of new technologies; centred on skill, risk, and hedonism; resistant to regulation and institutionalization; and ambiguous in their relationship to competition (Wheaton 2004: 12). These aspects of 'whizz' sports resonate with the thrills, spills and bellyaches of coastal kayaking that render it an activity that attracts adrenaline junkies and fitness enthusiasts. Kayakers share common interests and passion for getting involved in the sea beyond the perspective that can be gained from the land, and this passion helps form a common identity through their activities. As the following survey respondent puts it:

> Kayaking is the most enjoyable sport I have ever tried, I get to see places that I wouldn't normally see and it gets me out of my daily routine and in an environment where I have to focus completely and rely on my skills and those of people around me. It's one of the most sociable sports I've tried, the people I enjoy being around most are paddlers and I love being able to bump into someone [on the water] and be able to chat to them no problem at all.

Kayakers refer to the enjoyment and excitement of engaging in this practice, of 'the buzz of being able to do something active and exciting'. A crucial aspect of this enjoyment is the ability to be physically active; as a demanding activity which both requires and produces fit and healthy bodies, the sport 'is very relaxing, it keeps me fit, and I meet like-minded people who enjoy the exercise for fun', or as the following respondent puts it, 'it's fun, it's healthy, and it's something I feel successful at'. A further key dimension to this whizz sport is the camaraderie involved. Although not a team sport, sea kayaking is often undertaken in groups for safety reasons. Forty-six percent of kayakers who responded to the survey suggested that social reasons were central to their participation in the sport; as individuals put it, the 'main thing is the friendship of paddlers', 'I enjoy paddling with my friends, chatting and having a laugh is important while paddling'.

Sea kayaking is therefore experienced as an exciting, enjoyable and friendly pursuit by its participants, and as we will see below, this excitement is exacerbated by the risk involved in the activity. Such excitement (and risk) can be found in many sports, especially other whizz or extreme sports. However, as this chapter now develops, kayaking is also significantly different from this family of activities: there is something different about (being in) the sea.

The Sea as Viewed from a Kayak

> I hoped to ... come to terms, somehow, with the peculiar attraction that draws people to put themselves afloat on the deep, dark, indifferent, cold and frightening sea. (Raban 1999: 6)

'Yer in a kayak?"

'Yeah'.

'And yer going to paddle it around the South Island?'

I looked right back at him, smiled, and said, 'Ah yeah'.

There was a slight pause … He squinted his eyes as if he was trying to figure something out, then asked in a gravely puzzled voice, 'Why?'

How did I tell this guy..that I like the feel of the waves? … That I like feeling the sun and the wind and the power and the sensuous feel of ocean swells driving the boat forward …?' (Duff, to a fisherman on the subject of his paddling route around South Island, New Zealand 2003: 52)

Like Chris Duff (above), many individuals state they participate in kayaking due principally to the opportunity it gives them to engage with the sea. Participants articulate a strong affiliation between their sense of self and their belonging with the ocean; as one survey respondent put it, 'I feel most comfortable and at home on the water', whilst others express their attachment to the element itself: 'I love water and have an affiliation with it!' In order to find my own words for what this 'something' is about the sea, I went paddling.

I stand on the shore and look across the water. The sky is dappled blue. There are rain clouds in the distance, emptying themselves over England. I raise a smile at the lovely view. Carrying my boat to the water's edge puts up the gulls that were quietly swimming on the surface. I clamber into the cockpit.

The first thing you sense is your new orientation to the world. I'm now at 'ground' level. As adults, when do we ever see the world from this perspective? My familiar compass bearings become disoriented by this straightforward change in vantage point. Re-positioned to the land, I cast myself adrift from it with two simple strokes – left, right. How does this engagement with the sea change my senses? As I'm floating here a child's snow globe comes into my mind. On land, my life is set in such a hemisphere, and I am grounded, in the centre, at the bottom. The ground rarely moves, I take it for granted, and I have floating flakes above me. On the sea it is different. The hemi-sphere is wholed. My 'globe' is now a perfect sphere, partially filled with water, and I'm now floating in the middle, with a world around me. I become aware of the world of sky above, and the world of water below. Unlike the ground, the water beneath me isn't static. It's moving. *Although I know better, I feel it as if it were a body. I'm reminded of how Chris Duff calls the sea 'possessed' (2003: 214). It's definitely far from dormant. This space, as Tuan (1996: 445) would say, has 'personality'. The sea is lively, it's not a socially constructed metaphor, but it's here, it's present; if not quite a danger (yet), then active (Peters 2012). This morning the surface has small cats paws from the squall across the water, fractal mini waves on the surface, gathering into small waves, which will eventually become a series. Due to this surface movement, even when I do nothing, just sitting here with hands in the water, I move. The boat revolves to face the waves. They lap around me, slowing inching me backwards. I become aware of the easy but strengthening wind, and how I'm the weak link between these two elements. I'm the join between the sea and sky. My body could become a sail, my paddles too; catching the wind and moving me whether I want to or not.*

I know where these waves are coming from; I can see their source (the squall across the water) at the edge of my 'kayak globe' bubble. Yet there are forces acting on my water world that I can sense but I can't see. Soon the tide will turn. The second largest tidal range in the world will begin to act on my boat. Lunar power. Gravity. The whole damn universe is affecting me. And here I am in the middle of it all, on the water, at the mercy of deep mobilities I cannot fathom.

Sometimes you can see into the deep. When clear, the sea becomes a 'body' visibly alive with other bodies. Sometimes full of fish, sometimes saturated with sand, salt, even sewage, sometimes just too deep. Today it's a slurry brown. But I can imagine the rocks shelving out beneath me, and the forests of seaweed popping and gurgling in the submarine world. Above me the two dozen young gulls I put up earlier caw and wheel, curious about this interloper in the intertidal zone.

Although this new water world is strange and disorientating, many kayakers feel a sense of home on the sea. Respondents allude to how, from their point of view, it's the human condition to belong on the water. As the following respondent put it:

> Man [sic] has always wanted/needed to be close to water. For me it is more of a natural need. If I cannot be on it, I like to be near it. It gives me a sense of satisfaction.

Another kayaker states how their bond with water began at a young age: 'I've always been a "water baby"'; whilst another articulates how being close to water has been 'such an intrinsic activity to me through my youth that it just feels natural. I'm a kayaker and always will be.' Respondents here express how their human existence is a profoundly geographical – and oceanic – one. They are co-constituted by this watery realm. Kayakers, therefore, have particular spatial identities – they are co-constituted by water worlds. In the words of Webb, there is a sense of growing 'uncertainty about where the[ir] body ends and the rest of the [water] world begins' (2000: 3). This is perhaps unsurprising, as for those involved in kayaking activity their bodies are often partly submerged by water; as Sanford states, 'Paddlers emphasize their intimacy with the [sea] ... kayakers sit low in the water and are always partially immersed in the water' (2007: 882). I feel this intimacy strongly.

I sense each rise and fall of the sea through my boat and my body. I'm in the boat, and the boat is in the water. Importantly, it doesn't feel as if I am under the sea, or simply on the sea. The hull of the boat is a few inches under the water, and in the cockpit it is only this thin plastic hull that separates me from the deep. It's a very different feeling to being on a surfboard, or sitting on a surf- or wave-ski. They feel more like trays lying on top of the water. Kayaks have depth, have rails and edges; we are suspended in the meniscus between air and submergence. I feel this as I am now part of the boat. I'm wedged in, thighs, hips, feet. 'If your feet aren't bleeding', I was told when I first bought this boat, 'it doesn't fit right'. I tilt my hips, and the boat tilts in immediate response. I lift my left knee, I veer anticlockwise. I am semi-amphibious, a bi-elemental cyborg.

As an extreme sport, being positioned in this new medium is a risky venture. The kayaker is at the mercy of the elements and more aware of their vulnerability than they are on land. Duff is always conscious of this vulnerability, as he puts it:

> One of the challenges of sea kayaking is that every time the paddler snaps the spray deck in place and pushes off from shore, he or she is instantly in an environment that is potentially dangerous. Wind, tidal currents, reflecting waves, and water that is cold enough to lower the core temperature and kill by hypothermia are factors of the sea. (1999: 39)

This risk is something survey respondents acknowledge, tolerate and often become attracted to due to their feelings for the water:

> Kayaking has the added level of being an extreme sport with a level of risk that ensures you stay focused.

> I mountain bike and hike but never feel more alive than paddling on the edge in water, sounds a cliché, but it's true!

> I love being in, on, or around water. I enjoy being challenged physically and mentally and enjoy the environments that the sport takes me to.

Many respondents suggest that it is precisely the challenge presented by the water world that they enjoy. Participants enjoy the 'thrill of testing [themselves] against the water', whilst the 'challenge of you vs. the conditions, the planning, the problem solving, makes [your] heart pound' (CIWW Survey responses 2011). In a sense, these perspectives could be interpreted in ways that frame humans as against the water world, seeking to conquer it – as one respondent put it, 'I enjoy the feeling of mastering the elements' (see also Anderson 2010). However, many fall short of this urge to conquer and master nature, and appear satisfied by the combined mental and physical challenge offered by this element which they seek to move through safely and skilfully.

> I think it has a lot to do with the unknown and tackling the water knowing you can't control it but you can try your best to use it. Working out the water … is fun and speculative. (CIWW Survey response 2011)

Although risk is relative to age, skill, and perception (see Cloke and Perkins 1998, Standeven and de Knop 1999, and also Vannini, this volume, in relation to ice), kayaking does appear to offer participants a sense of risk – and safely practicing this activity in the face of this risk is vital to their sense of identity (Wheaton 2004).

A quarter of kayakers who responded to the CIWW survey stated that 'risk' is central to their motivation for kayaking, and as one individual states (perhaps with an element of dry humour), 'I kayak to cheat death and feel alive. I have an office job and knees are not up to jogging.' Speaking personally, this sense of risk and thrill occurs when experiencing the attack of coastal surf.

A vitally offshore wind. Surf. Sets of surf. Looming large from a distance, building walls of grey green, floomphing down in hard, fast collapse. The surf zone is a rough, encapsulated storm. Intimidating if I thought about it, so I don't. Speedily manoeuvre into the boat, get out there. The first rush of whitewater comes quickly, breaks the bow and explodes into me, bracing, invigorating, then gone. Shaking my head, salt piercing my eyes, a smile broadens and firecrackers of delight course my body.

From this level the waves are vast and vital. Limits are sensed, so through three broken sets I paddle then turn, prior to the rising battalions. I set myself for their boisterous collapsing. I attempt to gather pace, driving my paddle into the murk, but it's drawn back, being gathered into the beast. It comes, it's upon me. I lean back, feeling for the drive. The boat surprises; it's smooth, happy even, and with a pivot of my hips it turns easily along the lip, riding sideways and forwards towards the shore. Clean, fast. Hips pivot back, the boats wants to stay, but bracing my paddle it reluctantly straightens and instantly remembers the fun it can have being driven arrow-straight.

A dozen runs later, twists left and turns right have been nailed competently enough. And as I prepare my exit, a wave demolishes around me, dumping me onto the ocean floor, stealing my wind as down-payment and scouring my scalp into sand. Again I smile. Body stiff, torso aching, face weathered, I sit on the shore and see the tide rise, the beach withdraw and walls of sea build in magnitude and frequency. Braver souls than me continue the dance.

Coupled to the sense of risk in the water, participants also remark on the sense of escape they feel from engaging with the very different medium of the sea. Thirty-four percent of respondents choose to participate in kayaking as an 'escape', using their time on the water to leave behind personal and/or everyday problems and stresses. As the following survey respondents put it:

> I have always loved water and find water sports a release from real life.

> I get a sense of achievement whilst upping personal skills – something that is in contrast to my normal work environment. Being on the water is somewhere that is 'me' time. Being away from the city.

In these comments one may be tempted to suggest that kayakers emerge as others from their normal selves when they engage with water. In this element they are more at home, expressing themselves and their bodies in different ways to those articulated in their landed lives. In this new media, they can escape to and express a new element of their identity. Such a perspective would suggest that kayakers have a multiplicity of selves (Jameson 1991, Featherstone 1995), each articulated in different elements. Kayakers therefore have a spatial – perhaps even an elemental – *division* of identity (see Anderson 2004, 2012b). They can perhaps only be (part of) who they are when they are on the water. This co-constitution of self and water world is something sea kayaker Chris Duff has recognized:

> On the sea I was utterly myself. The wind, the seas, and the cliffs that resonated with the booming heartbeat of the ocean demanded that I put all other cares aside and be nothing but a paddler … What I feared was having to be something other than who I was on the ocean. (Duff, 1999: 222)

Whilst other sports may offer a sense of enjoyment, fitness and camaraderie, and other whizz sports may provide equivalent risk and challenge, it is only on water, and water encountered through the practice of kayaking, that participants can live out this crucial aspect of their identity.

So far in this chapter we have seen how the act of sea kayaking can be seen as a social, active, and adrenalin-fuelled practice. As an extreme sport it provides participants with a sense of thrill and risk, so much so that it enables individuals to articulate a different aspect of their identity to that performed on land. In the second half of the chapter we extend this idea of sea kayaking as an escape from land-centric modalities and consider it as a means through which participants can encounter a new way of viewing the world.

Water Thoughts

> Kayaking is like the initial plunge into a mountain stream, it refreshes and wakes the body and mind to new life. (Duff, 1999: 64)

With a few strokes I wheel around and try to navigate a 'path' parallel to shore. In considering my route as a 'path' I may be retaining my land vocabulary, but increasingly I'm having what Raban (1999: 40) refers to as 'water thoughts'. Duff talks of time being a strange thing when it 'slides past at the length of a paddle stroke' (1999: 87), and here I mark my own time by slicing the water with my paddles and splicing waves with my bow. This is the world at four miles an hour. At this speed, from this vantage point, I pay a different kind of attention. I see the small things – in same way that I do in a gallery when all is quiet and you have time to 'really look' (as Gormley suggests we should, 2009). I see swarms of midges over the water; there is still that 'urgh, get out of my face' reaction, but as I move through their suspension in the air, I sense the world not as a retrofitted movie existing on many differently layered spatial planes, but in perfect 3D interactive vision. Does the reflective capacity of the water and its distorted mimicry create this? Am I living in yet-to-be-invented surroundsensovision? This world must be here all the time but I simply fail to notice it. I now pay attention to the droplets falling from the rotating arcs of my paddle, how they fall in small ephemeral rainbow streams. I notice how the paddles delicately cut the sea surface, leaving circles that grow, fragment, and fade. I notice the water breaking off my bow, globules bending and sparks fizzing as I forge through the water. When I turn my head it's just beautiful: geometric fanning from stern to open ocean. Rhythmic fishtails.

The world at four miles an hour makes me notice the haiku of the moment. The small things, the interactions of – not letters, syllables, words and pauses – but paddle with water, flies with air. This overcomes me as interactions repeat, another and another, senses overload. I have to tune out. But then I tune back in again as it all filters back through me. I concentrate on just one thing. Smooth water over smooth rock. It flows like molten lava. Folding and mixing. It's as if I'm in a picture. But a picture that doesn't finish, but is constantly in the making. Can sea kayaking be considered as artistic performance? The beauty of it feels that way to me.

> every human being has the potential to be not only an observer of a picture but to be in the picture. That … matters. (Gormley, 2009: no page)

As Duff tells us, kayaking can be considered as a refresh and a reboot, not simply for our bodies, but for our minds too – sea kayaking can help us see the world anew. From my own experience, kayaking on the sea focuses my attention on the minutiae of life, so much so that it can render me as witness to and participant in something bordering on the artistic. Tim Ingold (1993) ventures into similar territory when he talks of 'taskscapes'. A 'task' in his view is 'any practical operation, carried out by a skilled agent in an environment' (1993: 158). The taskscape by extension is the 'entire ensemble of tasks [and] their mutual interlocking' (1993: 158). Ingold would argue that the act of performing these tasks and creating a taskscape can be seen as form of art, not in the sense that it becomes a definitive product or commodity that can be witnessed and sold, but as a process that is transient and ephemeral in nature. As he describes with respect to the task of painting,

> In many non-Western societies … what is essential is the act of painting itself, of which the products may be relatively short-lived – barely perceived before being erased or covered up. This is so, for example, among the Yolngu, an Aboriginal people of northern Australia, whose experience of finished paintings, according to their ethnographer, is limited to 'images fleetingly glimpsed through the corner of their eyes' (Morphy 1989: 26). The emphasis, here, is on painting as performance. Far from being the preparation of objects for future contemplation, it is an act of contemplation in itself. (Ingold 1993: 161)

This contemplation of the performance of the task, of a barely perceived act of creation that is both fleetingly lived and glimpsed, seems to me definitive of the practice of kayaking. The mutual interlocking of 'haiku upon haiku' feels to this paddler as an epic saga unfurling then disappearing through practice. As a solitary kayaker today, this art is for the maker alone, an audience of one. It does not matter that there is no permanent record of that beauty, or that no one else will see it. Its transience is perhaps part of its beauty. Its fluidity and ephemerality challenges my preoccupation by a world view dominated by the apparently solid and substantial terra firma. The sea's constant mutability offers me an alternative. The sea doesn't accumulate, congeal, or accrete. It doesn't sediment into conventional layers, or allow durable traces to be left in its surface. It is the counterpoint to the sculptured land which, for the artist Anthony Gormley, offers the opportunity for people to 'sense their own lives' in relation to something 'that doesn't move at the same speed that they feel they have to' (2009: no page). Rather than being like the sculpted land which offers a still point against which people can 'sense their lives', the sea offers a moving moment. The sea's mutability renders our lives longer and more permanent than a wave or storm squall, but its constancy renders that same life a fleeting ripple on its breaking surface. Sea kayaking makes us pay attention to these moments, offering us a reminder of the transience of all things and how this recognition is not something to rail against or resist, but is rather something beautiful.

Kayaking as Epistemology

> everything is fluid, even the land, it just flows at a very slow rate. (Goldsworthy 1994: 65)

The ability for sea kayaking to make us look at the world afresh, perhaps as a form of processual art, renders the pursuit more than an extreme sport or embodied identity, but as a new way of looking at the world. Sea kayaking produces water thoughts that critically challenge terrestrial assumptions of solidity, pace and permanence. In a similar way to Goldsworthy (above), it makes us aware that the world can be seen as continually in flow. Such new perspectives are so powerful that as I write this and consider the performance of art as something that lives and dies in the moment, any notion that we should seek to preserve experiences, congeal them into objects or accrete them into solidscapes seems anathema to the point of art itself. By seeking to capture and conserve a moment, it misses the point of creation and destruction entirely.[3] Sea kayaking therefore gives us a perspective which helps to recognize our involvement, vulnerability, and transience in the world. We begin to move away from seeing the world as

---

3   As Goldsworthy suggests in relation to his artistic construction of walls, he does not see these objects as solid, permanent edifices. In his words, 'The wall is not an object to be preserved in the traditional sense of art conservation. It is at the beginning of its life. What kind of life it has will depend on what happens to it. There are many possibilities ... Fragility and risk give the wall its energy. For it to retain this energy I must accept that the work has an uncertain future – even at the extreme of the wall being allowed to decay and trees left where they fall' (2001: no page).

'landscape' – as a 'congealed form of taskscape' (Ingold 1993: 162) – and every object as a 'collapsed act' (see Mead 1977) as these terrestrial terminologies fail to account for the fluidity of the water world. Water thoughts tell us that, as Inglis points out, scapes are 'living processes; [they] make men [sic]' (1977: 489). The scapes we live in are dynamic, and form our world views and our lives. Kayaking reminds us of the agency of water, the agency of the unseen, and the power of time, and their capacity to 'act back' on our lives 'in the process of their own dwelling' (after Ingold 1993: 163). When the practice of sea kayaking gives us these moments of reflection, it also gives us a new way of seeing the world. As Chris Duff recalls following a day of risk on the ocean:

> [Sometimes] this life on the sea, which at times seemed so difficult and vulnerable, seemed not to make sense. And yet I continued to return to it, perhaps because the sea demanded so much, but it also fed me in a way that no other lifestyle had. In its demands there was no masking the fragility of human life, yet in those extremes there was also a beauty that surpassed anything I had experienced. And in that intimacy of extremes, I had found the fullness and acceptance of who I was and how I fit into a very complicated and busy world ... The ever-changing, ravaging, and seductive sea forces one to acknowledge how tenuous and rich are the moments of a heartbeat, the flight of an albatross, and the arc of a rainbow. I closed my journal, not certain of but closer to understanding the power that the sea had over me. Maybe it was enough just to acknowledge the power of the sea and the value I placed in exploring my relationship with it. (2003: 242–3)

Conclusion

> being afloat gives me, at least, a heightened sense of being alive moment to moment. As small earthquakes do, it keeps you properly aware of your precariousness in the world. (Raban 1999: 90)

As the discipline of human geography tells its students, human are geographical beings. As we have seen in this chapter, some humans are formed, influenced, and feel most at home in water worlds. Kayakers enjoy lifestyles that are tied to moving water, connected through the emotions felt through immersing with it, and being mobile on it. Kayaking on the sea gives humans the opportunity to be other than who they are on land, and gives them a new way of looking at the world as a consequence. As the following kayakers put it:

> the world is more beautiful kayaking – it's why we do it, to interact with nature in a different way, from a different perspective.

paddling in natural environments is just good for my soul. (CIWW survey respondents 2011)

Kayakers are therefore one group of people who want to be challenged in relation to their constitution with the world (see Erskine and Anderson 2013 for others). Kayakers need *to move* and *be moved*, to be stimulated and challenged in terms of their relations to place. As spatial beings, this challenge comes from a geographical location itself; in this case a seascape's diversity and dynamism. Kayaking repositions humans in terms of their vulnerability and sensitivity and leaves them porous to new affects – of seconds of neat fear, but also moments of pure adrenalin and joy. To answer the question posed by the Kiwi fisherman to Chris Duff (above), accessing this new world view is what kayakers think about when they think about kayaking:

> Anyone who slips past those cliffs and carries on northward, enduring and exalting in the challenges of the passage, may well emerge from the experience with a different perspective, not in the sense of achieving anything of great notoriety but rather in the awareness of how one's life is a blessing, and if there is anything insignificant on the earth, it is the heartbeat of time in which we spend our lives. (Duff 2003: 257)

References

Anderson, J. 2004. The ties that bind? Self- and place-identity in environmental direct action. *Ethics, Place and Environment*, 7(1–2), 45–58.

Anderson, J. 2010. *Understanding Cultural Geography: Places and Traces*. London & New York: Routledge.

Anderson, J. 2012a. Relational places: The surfed wave as assemblage and convergence. *Environment and Planning D: Society & Space*, 30, 570–87.

Anderson, J. 2012b. Managing trade-offs in 'Ecotopia': Becoming green at the 'Centre for Alternative Technology'. *Transactions of the Institute of British Geographers*, 37(2), 212–25.

Brownley, R. 2010. *Cancer to Capricorn: The Path of the Modern Gypsy*. London: Reef Productions.

Cloke, P. and Perkins, H. 1998. Cracking the canyon with an awesome foursome: Representations of adventure tourism in New Zealand. *Environment and Planning D: Society & Space*, 16, 185–218.

Cooper, T. 2002. Up the creek, with a paddle. Daily Telegraph. [Online, 12 October] Available at: http://www.telegraph.co.uk/gardening/3305595/Up-the-creek-with-a-paddle.html [accessed May 2012]

Duff, C. 1999. *On Celtic Tides: One Man's Journey around Ireland by Sea Kayak*. New York: St Martin's Press.

Duff, C. 2003. *Southern Exposure: A Solo Sea Kayaking Journey around New Zealand's South Island.* Guilford: Falcon.

Erskine, K and Anderson, J. forthcoming. Tropophilia: A study of people, place and lifestyle travel. In *Travel and Transformation*, edited by G. Lean, R. Staiff, and Waterton, Farnham: Ashgate.

Featherstone, M. 1995. *Undoing Culture*. London: Sage.

Goldsworthy, A. 1994. *Stone.* London: Viking.

Goldsworthy, A. 2001 *Rivers and Tides. Working with Time: A Film by Thomas Riedelsheimer.* Artificial Eye. Germany / Scotland.

Gormley, A. 2009. *Making Space.* TV programme. More 4. 22 November.

Hanson, J. 2001. *Sea Kayaking: Outside Adventure Travel.* W. W. Norton: New York & London

Inglis, F. 1977. Nation and community: A landscape and its morality. *Sociological Review*, 25, 489–514.

Ingold, T. 1993. The temporality of the landscape. *World Archaeology*, 25(2) 152–74.

Jameson, F. 1991. *Postmodernism or the Cultural Logic of Late Capitalism.* London: Verso.

Jones, M. 2009. Phase space: Geography, relational thinking, and beyond. *Progress in Human Geography*, 33, 487–506.

Mead, G.H. 1977. [1938]. The process of mind in nature. In *George Herbert Mead on Social Psychology*, edited by A. Strauss. Chicago: University of Chicago Press, 85–111.

Melville, H. 1992. *Moby-Dick.* Wordsworth: Ware

Midol, N. 1993. Cultural dissents and technical innovations in the 'whiz' sports. *International Review for Sociology of Sport*, 28(1), 23–32.

Morphy, H. 1989. From dull to brilliant: The aesthetics of spiritual power among the Yolngu. *Man* (N.S.), 24, 21–40.

Murakami, H. 2009. *What I Talk about When I Talk about Running.* London: Vintage.

Peters, K. 2012. Manipulating material hydro-worlds: Rethinking human and more-than-human relationality through offshore radio piracy. *Environment and Planning A*, 44, 1241–54.

Raban, J. 1999. *Passage to Juneau. A Sea and its Meanings.* London: Picador.

Rowe, R. 2002. Foreword. In *Canoe and Kayak Handbook* by F, Ferrero. Bangor: British Canoe Union Pesda Press, 3.

Sanford, A.W. 2007. Pinned on karma rock: Whitewater kayaking as religious experience. *Journal of the American Academy of Religion*, 75(4), 875–95.

Sinclair, I. 2012. The Man in the Clearing. *London Review of Books*, 34(10) [Online, 24 May], 35–8. http://www.lrb.co.uk/v34/n10/iain-sinclair/the-man-in-the-clearing [accessed: June 2012]

Standeven, J and de Knop, P. 1999. Sport Tourism. Champaign: Human Kinetics.

Thoreau, H.D. 1864. *The Maine Woods.* Boston: Ticknor & Fields.

Tomlinson, A., Ravenscroft, N., Wheaton, B. and Gilchrist, P. 2005. *Lifestyle Sports and National Sport Policy: An Agenda for Research.* Brighton: University of Brighton.

Tuan, Y-F. 1996. Space and place: Humanistic perspective. In *Human Geography. An Essential Anthology*, edited by Agnew, J., Livingstone, D. and Rogers, A. Oxford: Blackwell, 444–57.

Van Bottenburg, M and Salome, L. 2010. The indoorisation of outdoor sports: An exploration of the rise of lifestyle sports in artificial settings. *Leisure Studies*, 29(2), 143–60

Walls, M. 2012. Kayak games and hunting enskilment: An archaeological consideration of sports and the situated learning of technical skills. *World Archaeology*, 44(2), 175–88.

Webb, J. 2000. *Beaches and Bodies*. Available at: http://people.brunel.ac.uk/bst/documents/jenwebb.doc [accessed: May 2012].

Wheaton, B. 2004. Introduction: Mapping the lifestyle sportscape. In *Understanding Lifestyle Sports: Consumption, Identity, and Difference*, edited by Wheaton, B. London & New York: Routledge, 1–28.

Chapter 8
Deep Ethnography:
Witnessing the Ghosts of SS *Thistlegorm*

Stephanie Merchant

Introduction

The majority of the contributions to this book have concentrated on the practices and performances being carried out above the water's surface. Like landscapes, in these chapters the sea is constructed as something to be 'on'. By contrast this chapter explores the three-dimensionality of ocean space by delving beneath the water's surface and highlighting the ways in which ocean space is moved *through*. To do so, I briefly re-tell the story of SS *Thistlegorm*, a 420-foot casualty of the Second World War. Today the ship rests at the bottom of the Red Sea forming arguably the most structurally and content rich wreck dive site in the world.

If ships have stayed largely beyond the remit of geographical interest (as Laloë notes, in this collection), then it follows that embodied research on shipwrecks fails to even register as a conceivable topic of inquiry. In this chapter, through an auto-ethnography of the SS *Thistlegorm* wreck site, I attempt to demonstrate that the over and done with materialities of shipwrecks are worthy of discussion and pertinent to contemporary geographical debates. Thinking through notions of haunting and materiality, I argue that the objects and beings found at wrecks are so situationally transgressive that their competing material signification (in the form of decay, biological take-over and their potential to evoke intertextual memories), has a noteworthy influence on the embodied practices and performances being played out by those who visit the site. In doing so, I contend that in unpacking ocean experiences, we need to look beyond the boundaries of the ocean surface, to ocean depths.

Engaging with the concept of 'hybridity', geographers have begun to write of the differing embodied affects that are felt and produced by those who experience sites of obvious tension (in terms of the meanings attributed to the things and atmospheres which furnish them). Whether it is the tension between past and present, the real and the imagined or the natural and the cultural, ghostly and haunted environments have emerged as popular topics of academic research. Humans connecting, feeling or being affected by intangible others/things/bodies throughout their daily and otherwise routines have been studied notably by academics such as Edensor (2005a, 2005b, 2005c, 2007), Gordon (1997) and Bell (1997). In this chapter I open up these geographies to the underwater spaces of shipwrecks in order to consider the ways in which history comes alive, is (re)performed and is distorted during encounters with sunken artefacts, colonized passageways and alienating environments. I do so by firstly outlining the SS *Thistlegorm*'s late history. Following this I explore the concept of hybridity, to consider the ship's re-birth as a natural-cultural ecosystem and also in terms of temporality, to consider how 'ghosts of the past' infiltrate experience of the present. Before concluding, I present an autoethnography which details my journey of exploration from dive boat to sunken ship and back, twice over.

Throughout the chapter I highlight that it is possible to take contemporary, land-based geographical debates of materiality, haunting, memory and place meaning to the ocean depths, and to consider the shipwreck as the knot which is the means by, and the point at which, these themes come together. Consequently, SS *Thistlegorm* is seen to be an 'other space' where nature-culture, past-present boundaries are visibly in flux, where history is brought back into the present and relived through interactions with encrusted artefacts. The chapter is presented as a two dive tour of the ship in terms of these two binaries. Thus the description deals with evocative materialities as well as considering the simultaneous decay and biological takeover of the ship. In other words, it attends to the ways experience is entangled with materialities, and where experiences in relation to those materialities are often transgressive because of the underwater setting. However, although these two themes are presented as distinct moments of obvious connection (with the first dive largely concentrating on nature-culture and the second past-present), they cannot be thought of in isolation.

A History of SS *Thistlegorm*

SS *Thistlegorm* is a sunken World War two cargo ship that lies beneath the Red Sea, between the reef of Sha'ab Ali and the Sinai Coastline. For divers visiting Hurghada or Sharm el Sheikh, *Thistlegorm's* structural remains are one of the most popular attractions. Resting at 30 metres below sea level on a sandy bed, SS *Thistlegorm* is accessible to most divers and the destruction of a large section of its superstructure makes for novel routes of exploration. At peak times the wreck can have up to 30 dive boats moored to it on a daily basis and overcrowding below water is common.

With a length of 126 metres, a 17 and a half metre beam and a capacity of 4,898 tonnes, Ecott (2007) argues that SS *Thistlegorm* provides one of the most exciting wreck dives in the world. Constructed in 1940, the freighter was destined for Alexandria to supply British troops with equipment. The equipment ranged from small scale items such as Wellington boots, medical supplies and ammunition, to large scale vehicles including two locomotives, complete with accompanying cars, as well as tanks, trucks and motorbikes.

Having set out from Glasgow, SS *Thistlegorm* had made the long journey south, round the Cape of Good Hope and back north up the West African coast, to the Straights of Gubal. As the ship lay at anchor on 6 of October 1941, in the so called 'safe zone', it was targeted by a long range German bomber that had travelled from Crete. Struck twice towards the rear of the ship, the two bombs set off a subsequent explosion of the on-board ammunitions located in hold number four. The stern was blown apart and nine seamen perished in the incident, with the remainder of the crew escaping to the nearby SS *Carlisle*. John Kean's (2009) book devoted to the history of SS *Thistlegorm* tells the story of the ship and its crew members from its construction in Sunderland to the night of its demise the following year.

Fifteen years after the event, with the tragedies of the Second World War over but not forgotten, Jacques Cousteau and his team of 'underwater scientists' set out from Marseille on a wreck seeking expedition. Travelling south through the Gulf of Suez, he and Frederic Dumas flagged up SS *Thistlegorm* on a nautical chart, believing it to be a promising find. Following a preliminary dive by Dumas and Falco, Cousteau decided to return to SS *Thistlegorm* for further exploration. They moored the Calypso up at the wreck for two days to film its structure and contents, with the footage subsequently featuring in the Palm d'Or winning documentary *The Silent World* (1956), and the 1963 book *The Living Sea*. Cousteau's videography and literary descriptions provided information on the ship's whereabouts and structural quality for all to explore.

Despite the success of Cousteau's dives and subsequent publicity, in the 1960s and 70s the SS *Thistlegorm* failed to attract divers due in part to recreational scuba remaining in its infancy. It was not until the early 1990s that SS *Thistlegorm* began to attract significant numbers of tourists, when the towns of Sharm El Sheikh and Hurghada, on opposing coasts of the Red Sea, began to develop into Egypt's most popular diving resorts.

Although an attractive dive site, the official status of SS *Thistlegorm* remains controversial. It is unclear whether 'accidentally' sunken ships can be classified as artificial reefs (as opposed to the strategic and deliberate sinking of ships), and as such SS *Thistlegorm* would probably not 'technically' fall under this category. Despite this, the characteristics of deliberate and accidentally sunk wreck sites are often similar (Stolk et al. 2007). Shipwrecks occupy the benthic zone of the marine environment (Stolk et al. 2007), are capable of supporting wildlife populations, and they are often characterized as being 'artificial habitats'. Irrespective of the 'naturalness' of the site, academic research carried out on the great lakes of America has highlighted that divers and tourists are more keen to visit shipwrecks that *weren't* deliberately sunk because of the added historical interest associated with them (Vrana and Halsey 1992). In this sense, these sites are a form of 'modified space' (after Lawton and Weaver 2001) where 'a recreational user can experience a combination of social and natural history', 'a novel setting for humans to interact with natural environments, and … cultural heritage' (Stolk et al. 2007: 336).

Having outlined the history of SS *Thistlegorm*, the chapter now turns to recent academic literature on haunting and materiality in order to contextualize my motivations for highlighting the ship as a potentially interesting research site for unlocking the underwater world.

Geographies of Haunting

Peters (2010: 1267) recently stated that the liminal spaces of ships and the sea 'may well be sites that are specially suited to investigations concerning spectrality and magic'. Further, she argues that studying ships in this light would offer possibilities to consider the 'folding and blending of past and present in material manifestations'. In this section I span the literature of geography's dealings with spectrality, hauntings and ghosts, or what has been called 'spectro-geographies' (Maddern and Adey, 2008).

Bell's (1997) article 'The Ghosts of Place' was one of the first to deal exclusively with the connections between ghostly beings and their residence in certain places. Whether they be friendly or threatening, his varied examples illustrate that ghosts of place are at once highly subjective yet situationally specific 'beings', whose presence tells us a lot about the ways we feel and relate to certain spaces. Ultimately Bell describes ghosts as the haunting manifestations of social experience of place, and rather than 'being given to us' (Bell 1997: 830), ghosts are projections that *we* populate our surroundings with.

Since Bell's publication, ghosts have figured in a variety of guises within geography, with some being more loyal to their popular culture configurations than others. Most however, have been to varying degrees conceptualized as mysterious affects (Holloway 2006, Wylie 2005, 2007), sensations (Edensor 2007, Pinder 2001), distorted acts of memory (Bell 1997, Crang and Travlou 2001), or the spectral traces of an author (Derrida 1994, Wylie 2007). Furthermore, whilst some consider ghosts within the everyday spaces of the city or the home (Bell 1997, Pile 2005), others have concentrated on more extra-ordinary spaces. For example, for Gordon (1997: 64), ghosts are 'haunting reminders of lingering trouble', and as a consequence, hauntings are often associated with sites where acts of extreme wrongdoing such as slavery or torture have occurred. Other less routinely experienced hauntings include those of tourist sites (Bell 1997, DeLyser 1999), derelict ruins (Edensor 2005a), or the sites of religious/spiritual encounter (Holloway 2006, Holloway and Kneale 2008).

Alongside this breadth of ghostly manifestations and locations there has been an increasingly emergent body of literature that puts materiality into the 'spectro-geography' mix. Tim Edensor's work in this sub-genre is particularly prolific (2005a, 2005b, 2005c, 2007). Edensor argues that interacting with the over and done with architecture of industrial ruins and their varied scattered and defunct content, encourages a human comportment that is free from the standardized and regulatory framework of city space. This 'alternative sensual realm' allows the body to perceive in a more holistic and richly experiential manner. Such experience he argues, can 'conjure up the forgotten ghosts of those who were consigned to the past' (Edensor 2005c: 311). DeSilvey is similarly concerned with the potential decayed and rotting materialities have as intertextual memory aids (2003, 2006, 2007). Though not dealing explicitly with the concept of haunting, DeSilvey does argue that historical artefacts complicate our experience of the present by bringing forth memories and sensations associated with the past (2006).

It is these latter two authors whose work I believe is of most relevance to this chapter, largely because the figure of the ghost in these is no more than an unexpected recollection of an old memory or a musing of relations instigated by the touch of a semi-decayed and forgotten object. As Holloway and Kneale have noted, 'what we are dealing with when a space becomes haunted, is the disruption or dislocation of normalized configurations and affordances of materiality, embodiment and space' (2008: 302). Thus, in this chapter I see hauntings as the uncanny moments of wonder and confusion, the random and complex mental and sensual associations that emerge as a result of interaction with the structure and contents of SS *Thistlegorm*. The 'tangle of temporal linearity' is what the chapter aims to explore (Maddern and Adey 2008: 292), the intermingling past and present, real and imagined, absent and present *in the present*. It is for this reason Bergson is owed a mention here. In *Matter and Memory*, Bergson states;

> There is no perception which is not full of memories. With the immediate present data of our senses we mingle a thousand details out of our past experience. In most cases these memories supplant our actual perceptions, of which we then retain only a few hints, thus using them merely as signs that recall to us former images. The convenience and rapidity of perception are bought at this price; but hence also springs every kind of illusion. (2004: 24)

Although Bergson's 'details' seem more micro-scale than some of the recollections and affects I explore, with regard to SS *Thistlegorm* his theorization of memory remains relevant. During the time I was diving, my perception of SS *Thistlegorm* was impregnated with often barely relevant recollections, reinforcing the subjective nature of my engagements with the ship space.

> In short, memory ... covering as it does with a cloak of recollections a core of immediate perception, and also contracting a number of external moments into a single internal moment, constitutes the principle share of individual consciousness in perception, the subjective side of the knowledge of things. (Bergson 2004: 25)

Thus temporalities 'collide and merge' in a seascape of juxtaposed 'asynchronous moments' (Edensor 2005c: 324). Furthermore, it is not just SS *Thistlegorm*'s temporal associations that call forth these obscure traces of memory and association, but also its transgression from the human world, its semi-natural, semi- manmade structure and its partial destruction and disarray, elicit a confusion and enhanced mode of sensory perception. Framed thus, we can begin to see the mixing in of two bodies of literature that have at their core a shared interest in dispelling fixity and singularity of meaning.

The Journey

After equipment assembly and a three hour windy journey on the 'fast' boat, we arrived at the site and began transforming into human-fish. First I turned on the airflow on my tank and began to don a rented, tired, five millimetre wetsuit. Then the weight belt; I tend to be 'floaty' towards the ends of dives and the high salinity of the Red Rea makes this even worse. The four kilos of lead strapped round my waist to counter this dug deep into my hips and as I simultaneously swung my spare arm into my buoyancy jacket and stood up, the weight of my tank pushed the lead further into my spine. Drastically top heavy, my thighs burnt as I bent over to pick up my fins and mask from equipment box number 8 (my designated box for the week). Once all strapped in and buckled up, I checked my 'buddy's' equipment – all was fine, as was mine. I took a drag from each of my regulators, just to make sure, and on cue the captain blew the ship's horn- the boat was securely anchored and it was safe to jump in. Laden with equipment, we were the first pair to hobble to the back of the boat and as we took it in turns to slip on our fins and spit in our masks the sea rose and fell, slamming against the dive platform and testing our balance. Leading the way, one by one we took giant strides off the platform and gave the 'ok' sign to the deck staff. Seconds later the Red Sea drew each prostheticized body under (Merleau-Ponty 1962).

Dive 1

Leaving behind my land-based habitus (Merchant 2011b) and following an anchor line into the depths, the remnants of a ship became visible. Eerie, dark and monochrome, it was not an inviting sight. The entirety of its scale was not taken in at first, for it took a good minute or two to adjust to the environment. The rising mid-morning sun largely failed to penetrate the deep; the sea's turbid waters scattered and refracted its rays, denying any assurance to us of what may lie ahead.

   Sinking, further and further to the sandy bottom (30 metres below sea level) insignificance in the face of magnitude was realized. Against a backdrop of nothingness, *Thistlegorm's* form trailed endlessly off into a murky horizon, where visual acuity was no longer and where structures merged into little more than dark blocks of grey set against a deep blue. Breathing was an effort and the turbulent stream of bubbles I exhaled rattled the regulator against my gums and noisily massaged the side of my face. Save my 'buddy', the others had dispersed now, searching for isolation and escape from the traces of mass tourist exploration that this dive site has come to be subjected to (Kean 2009). Kneeling on the sea bed amongst the metal debris, the wreck appeared void of any life. No longer within its intended realm, SS *Thistlegorm* felt deserted, a melancholy reminder of the loss and destruction of the Second World War.

Orientated and inquisitive, we made our way to the rear of the ship, passing over the hull. Rubble like in appearance, the matter that fans from the scar of the ship is ambiguous. Hold number four, the impact zone of two Luftwaffe bombs which subsequently caused the explosion of ammunition stored on board, 'is almost unrecognisable as a ship'(Cousteau 1963: 87). Here the inside spills out and the outside has been washed in, mangled and deformed, 'the mutilated plates ... twisted and rippled like kelp'(Cousteau 1963: 87).

Spilling from the scar, beneath the chassis of a jeep, a box of four inch shells could be seen. One had previously been scrubbed clean by a diver and through a thin veil of dark green slime, the engravings were easily visible. I didn't bother to translate the Roman numerals, but was surprised by the date inscription: '1929'. The shell box is but one of many relics of SS *Thistlegorm's* dense and disorganized 'temporal collage' (Lynch 1972), encompassing materials from the 1920's to the occasional contemporary discarded fin or torch.

My attention turned to focusing my digital camera on the shells and to maintaining buoyancy. As the captured image presented itself on the camera's digital screen, I was reminded that the ship was not as dull as it seemed. The absent colours became momentarily present through the red filter on the lens; allowing rust, algae and coral to emerge as distinct colonizers of the steel remains, no longer subsumed together by their similar crusty texture, but more akin to DeSilvey's 'technicolour moulds' (2006: 319).

I felt the cold flush out the arms of my ill-fitting wetsuit. Although the water was a temperate 19 degrees, a shudder took over my back and shoulders. Underwater the body loses heat faster than in air (Adolfson and Berghage 1974), and it was not long before I felt the chill deep in my bones. Unlike reef dives, SS *Thistlegorm* has a different aura. Darker, deeper, and despite being manmade it is an alienating place.

Swimming anti-clockwise around the outside of the wreck the dominant structure of the anti-aircraft cannon came into view. Here the ship leans so precariously it appears to be held upright by a single rope from above. In line with the ship the cannon has swung round to lean portside, aiming ominously at the sandy bottom bellow.

Looking up the barrel, I had been told, makes for a well-composed photo. But as a newcomer to the art of underwater photography, my task ran far from smoothly. Like most of the ship's exterior, my subject was covered in coral, and fish life was more abundant. Their silhouettes scattered back and forth in the background, shying away as I moved and returning as I became still. Despite its intended use, this tool for combat, death and destruction has become 'converted to peace' (Cousteau 1963: 87), the foundation upon which an ecosystem has formed, the residence of bat fish, turtles and crustaceans profiting from the 'procreative power of decay' (Bataille in DeSilvey 2006: 320). After three hovering and flashless attempts to capture the scene, I remained steady enough to succeed.

Rounding the stern of the ship the propeller juts out from the sand. The two emergent blades were taller than me and both were covered with a patchwork of algae and coral, themselves at war over space, they camouflaged its surface merging with the sandy bottom, making it difficult to see where the metal ended and the sea bed began. A lone star fish lay protected from the current beneath the fin roof and as I turned to point 'him' out to my buddy a school of bat fish hurried past and disappeared off to the shelter of the ship's hull.

As Edensor argues, beyond the remit of 'social circulation', the ocean's processual and ecological rhythms highlight that human made objects don't solely have social lives, but chemical and biological lives too (2005a: 100). Whilst fish swirled around the propeller's blades as though they were 'waves of the sea', the sharp streamlined edges of steel have been softened by filaments and fronds (Cousteau 1963: 87) and corroded by salty water. Thus, the marine life seems to offer, a 'future possibility, a hope' that along with tourist interest gives SS *Thistlegorm* purpose again (Gordon 1997: 64).

As I finally became accustomed to the awesome surroundings, I felt another trickle of cold run up my arm and travel round my wetsuit, lighting up its failure (Heidegger 1962). Breathing was still a strain and the visibility began to diminish seemingly by the second. All these aspects acted as a reminder that SS *Thistlegorm* has transgressed from the human world, seeming ironic that above the surface, these characteristics would likely have a negative impact on the experience. Underwater though, these are what make exploration unique, interesting and undoubtedly popular; the cold, the physical effort, the required bodily technological mediations and prosthetics, the absorption of colour and light, the distortion of sound, the erosion, demise and colonization of its form. Had the ship not sunk, had it not transgressed the surface it wouldn't be nearly as interesting, enchanting, chilling (literally and metaphorically), structurally and atmospherically. These elements are what make for adventure, but they simultaneously and paradoxically are what detract from it.

I checked my air gauge for the first time; only 60 bar remaining,[1] but only 35 minutes down. Normally an efficient breather, the depth and overbearing surroundings must have made me gasp. We fought the current, paddling with force. With the visibility worsening our immediate surroundings seemed not to change – a homogenous superstructure to the left and featureless blue to the right – it was difficult to tell if we were even moving forward. No longer inquisitive toward the relics, my attention had turned to seek out our anchor line, tied somewhere to the middle of the ship. Finally and with less than the recommended air left to ascend we latched onto the target for the final few minutes of the dive. I looked down upon 'the petrified ship' at only five metres below sea level (Cousteau 1963: 85). Despite my previous scepticism of such oft quoted remarks, for the first time as a diver, I felt as though I was flying. It's one thing to look down upon a reef, another

---

1   This constitutes roughly 1/8th of the tank. Divers normally begin ascent when they have 50 bar remaining.

to see a ship's contours from above. That said, SS *Thistlegorm*'s scale is so vast that it doesn't invite a static contemplative gaze, the bow and stern were beyond perception and from that distance neither my torch nor the camera's filters could recover the colours lost to the sea.

Dive 2

Much shallower this time at around 18 metres, we sank to SS *Thistlegorm*'s deck. There's a large opening in the floor where the hatch cover has been blown away. Sheltered beneath the remaining sections of deck, rests a selection of 1940s vehicles. A motorbike stands as if on display, just like the one Steve McQueen rode to flee the prisoner of war camp in The Great Escape – probably worth a fortune had it not suffered its fate. Somewhere on land there is an identical one surfaced by Jacques Cousteau, along with the captain's safe and the ship's bell that chimes eerily in *The Silent World*, accompanied by a female voice that whispers its inscription of '*Thistlegoooorm*' (Cousteau and Malle 1956). These recollections, some of real events and some of fiction, bring forth an 'imaginary past', making SS *Thistlegorm* seem uncanny, the site at which 'temporality and spatiality collapse' (Vidler, 1999 in Edensor 2005a: 836) .

Circling below deck not far from the bike is a jeep, windscreen intact. A female diver, sat at the wheel. I recognized the luminous pink fins – it was the excitable American woman from 'my' dive boat. I wasn't sure if I approved of her mimicking practices. It seemed simultaneously disrespectful yet a good photo opportunity. Here it can be seen that the ruination of SS *Thistlegorm* presents a confusion over, and absence of, boundaries. Slipping beneath the ocean's surface, norms of comportment become altered yet can jar when juxtaposed to transgressed materialities. The plethora of possible lines of action afforded by the wreck seems to encourage a human comportment counter to that traditionally expected in a land-based heritage site. As Edensor notes, 'in ruins [one] becomes strangely reminiscent of childhood sensory immersion and of the pleasurable negotiation of space largely denied to adults' (2005a: 837). The 'look but don't touch' ethic is replaced by a playful urge to act out or mimic ship performances, 'conveying a sense of corporeal empathy' towards those passed (Edensor 2005a: 840), bringing the rusty and encrusted materialities of the ship temporarily back to their original use, no longer banal but remarkable in their contemporary transformed state and site.

I decided not to copy the female diver and instead returned to the motorcycle – the others had moved on and there was enough time to take a photo. Between dives the flash had become jammed, and whilst this negated the need for the colour filter, the light from the flash now bounced off the plankton and the floating matter, previously barely perceptible in itself, it considerably reduced my visibility of the subject.

A testament to my state of jumpy unease, I felt my knee 'land' on something. I looked down and inhaled from shock, then instantly became calm as the moray eel I thought I'd made contact with, became a rubber tire emerging from 'years of sea dust' (Cousteau 1963: 80). Prehistoric looking, with their dominant jaws, beady eyes and pointy teeth – SS *Thistlegorm* seemed right for eels. Cavernous, dark, and far from the human world, the perfect place for the forgotten and unsightly, or as Cousteau describes; 'the rats of this forsaken garage'(Cousteau 1963: 85).

As I let the remaining air out of my BCD and inhaled, my buddy and I both rose 'through' the deck to the captain's bridge. On my way up I spotted a tank resting on its side, so 'thickly dressed' with *Ectopotra* it had almost lost all contours and was recognizable solely by its distinctive wheel and track design. The bridge itself no longer conformed to the conventions of tidiness. It represents of a kind of loss, of life and unfulfilled wartime duty. It was here that I thought of people, of sailors, of workmen and the captain. The rest of the ship didn't feel lived in in the same way, more for storage and transport than being, walking, and working. My buddy and I kept rising further up over the deck's ladder and ventured right, off the side of the ship. We swam through an opening to the captain's quarters, moving amongst 'the presence of those who are not physically there' (Bell 1997: 813). We entered the empty chamber, gutted of all remains, far less exciting than Cousteau's (1963) descriptions in *The Living Sea*.

Over the years the litter of corked bottles and porcelain crockery has been filched by trophy seekers, wanting material evidence to prove their exploratory prowess. Such objects, handled by the sailors, are imbued with an 'aura' (Benjamin 1999), irreplaceable, along with the ship itself; they are the means by which we interact with the past. To the left a doorway led us to the bathroom which contained only a bath. Although minimal in its furnishings it was not difficult to imagine what went on there. I recalled movie footage of my father's 'World at War' DVDs and passages from Kean's (2009) historical accounts. Somewhere near the bridge a heroic sailor saved another by running barefoot across the burning metal deck, receiving the George Cross for his efforts. As DeLyser says of Bodie – California's surviving ghost town – discovering SS *Thistlegorm* is an 'intertextual experience' (1999). For me, the ghosts of SS *Thistlegorm* are not just the traces of those who lived and worked on her, but also those who dwelt in my memory, and who were brought to being through my interaction and exploration with/of the ship. As Gordon explains; ghosts are social figures, 'of inarticulate experiences, of symptoms and screen memories, of spiralling affects, of more than one story at a time' (1997: 25). The captain's bridge then, 'evokes the process of remembering itself, its impossibilities and its multiplicities' (Edensor 2005a: 834).

My air gauge told me I was halfway through my time so we moved out and on to the nose of the ship. I spotted 'Pink Fins' posing again, this time acting out a flying Kate Winslet, from that well-known scene in *Titanic*. I wondered if she was singing along to Celine Dion's accompaniment of *My Heart Will Go On* in her head. She overtly exemplified Edensor's (2005a: 834) argument that 'the qualities or affordances of particular kinds of space, those full of random juxtapositions, clutter, obstacles, and numerous pathways, [also] demand a fuller performative, corporeal engagement with space and hence with memory.' The ghostly animation of material things by means of the elusive and overlapping perceptual information they bring to attention in the diver exemplify our multiple and tenuous relations to 'clutter'; rather than the clutter itself, they are what produce 'an excess of meaning, a plenitude of fragmented stories, elisions, fantasies, inexplicable objects, and possible events' (Edensor 2005a: 834).

Back over to the starboard side, there's a semi-covered passageway. We travelled part the way along and emerging from a bundle of unidentifiable matter we spotted a clown fish family nested deep within a blanket of *Heteractis Magnifica* or, 'The Magnificent Sea Anemone'. Both flora and fauna were brightly coloured under my torch's gaze and stood in stark contrast to their dim surrounds, almost iridescent their hideaway formed 'an impressive display of animal [and plant] adaptation to available resources' (DeSilvey 2006: 322). Since learning to dive I must have seen hundreds of clown fish and at the risk of dangerous anthropomorphisms they seem to represent the sedentary nuclear family of the fish world. Concerned (but not too concerned) that they probably didn't appreciate flashing lights, I only took a couple of photos. The miniature fish hid deep in the anemone's tentacles, and the 'parents' swam close by to protect 'him' – retreating when I thrust the strobing plastic box in their faces. Neither 'artefact nor ecofact, but somewhere in between', SS *Thistlegorm*, as a home to such creatures, has begun to take on an 'ecological function'(DeSilvey 2006: 323), a 'blurred terrain where nature and culture are not so easily (as if they ever were) distinguished and dichotomised' (Harrison et al. 2004: 9).

Harassing over, we propelled our way back into the ship's interior, through an opening in its side. It could have been a window or a door, I didn't notice and it seemed irrelevant, after all, 'processes of decay and the obscure agencies of intrusive humans and non-humans transform the familiar material world' (Edensor 2005c: 318). This section of the ship was empty too, just a maze of rooms which have acquired additional entrances/exits and the disorder brought on by a major explosion. 'Wires [hung] in garlands from overhead' forming a maze-like selection of passageways for the equipment laden diver who is clumsy and bestowed with an oversized turning circle (Cousteau 1963: 86). Whilst SS *Thistlegorm* does chime with Edensor's notion of ruins allowing for reminiscence into childhood sensory immersion, there is still a hint of adult/rational caution to be had, in my mind at least. Cables and metal edges drop down from the ceilings, accompanied by snake-like chorals and sponges. I checked my gauges once more and became wary of entangling my second stage (or regulator hose). As a slight claustrophobic above sea, the thought of becoming ensnared or blocked in with a limited air supply made me apprehensive. To make sure this didn't happen I became much more aware of my positioning in relation to the surroundings, and whilst compensating for an impoverished sense of proprioception I looked up to check my clearance. In doing so I noticed that the water was glimmering on the ceiling of the chamber – 'a quicksilver mirror that distorted our reflections' (Cousteau 1963: 86). I raised my hand and it passed through the mirror into air, the exhalations of those who had been there before me. Holes have been drilled all over the ship's body to prevent such build ups, and from the exterior, bubble trails could be seen mushrooming and morphing almost endlessly from wreck to surface, set against a background of dancing rays of light. Into the next chamber and out through the roof, *clang*, my tank chimed and vibrated as I collided three times with the edge before judging my width correctly. A deep breath, I could relax. The route to the surface was clear should anything happen.

Before ascending back to the dive boat we looked back to the deck. Amazingly a complete train engine has held its position on the starboard side and beside it amongst the crumpled metal, the engine's cars lie deformed and imploded under the weight of the ocean.

Conclusion

In this chapter I have sought to descriptively explore a particular underwater environment, the once human- and surface-based realm of SS *Thistlegorm*, in its current resting place 30m below sea level. Through witnessing, capturing, touching and interacting with the ship, its materialities and its avenues of exploration I have begun to demonstrate that places beyond normal remits of exploration, in this case below the ocean's surface, can prove interesting and useful as geographic field sites. SS *Thistlegorm* presents a vast array of research possibilities, and here I have chosen to touch upon two key themes. As a displaced remnant of the Second World War, SS *Thistlegorm* illustrates that objects/vessels/buildings have natural as well as social lives. SS *Thistlegorm* is currently becoming re-humanized by dive tourism, but it is simultaneously becoming wilder by the minute as corals grow and as fish reside at the same time as the ship's cultural information is scattered, filched and colonized by the sea's currents, materialistic divers and homeless species. The ships' architecture and its previous uses interact with its transgressive occupation of the sea bed and its inhabitation by various forms of flora and fauna, such that stairs, baths and cupboards become doubly coded as sites of decay and colonization. Thus, a sense of loss or nostalgia for times gone bye is fostered, yet equally as the dwelling places of crustaceans, corals, eels and their residues, the barnacles and worm engravings become 'traced out on the material textures of ruination'(Edensor 2005a: 842) and highlight a 'matrix of memory'(Casey 2000: 311).

The character of this sub-aquatic environment is constructed as chilling and eerie by a combination of the sunken ship's broken and elderly materiality, its signification as a reminder of war, as well as the characteristics of the water itself; not just dark and colourless, but physically cold, deeply pressurized, full of scattered particles as well as a strong current. Thus I have employed SS *Thistlegorm* to consider the 'density of experience'(Adorno in Gordon 1997: 240) and the multiple and overlapping politics of space being played out there. Through interacting with the site's materiality, visually and performatively I have explored the potential the sea's depths and in particular wreck sites have for re-animating the past and considered the ways in which real-time perception is impregnated with it (Bergson 2004: 24). Perceptions involve 'an effort of memory' that highlights the mingling of previous images. For example my (non) site specific recollections of Cousteau's filmed dives of SS *Thistlegorm*, general wartime documentary footage, literary articulations of shipwrecks, with perception of the present, bringing 'together all sensible qualities', in an enriching experience yet one that is vastly subjective in its knowledge creation of material things and environments (Bergson 2004: 48). As Bergson claims, since perception is inseparable from memory, the past is imported into the present, such that we are constantly driving toward the future, and thus de facto forcing us to 'perceive matter in ourselves'(2004: 80).

References

Adolfson, J. and Berghage, T. 1974. *Perception and Performance under Water*. London: Wiley.

Bell, M.M. 1997. The ghosts of Place. *Theory and Society*, 26(6), 813–36.

Benjamin, W. 1999. *Illuminations*. London: Pimlico.

Bergson, H. 2004. *Matter and Memory*. London: Macmillan.

Casey, E.S. 2000. *Remembering: A Phenomenological Study*. Bloomington: Indianna University Press.

Cousteau, J. 1963. *The Living Sea*. Middlesex: Penguin.

Cousteau, J. and Malle, L. 1956. *Le Monde Du Silence*. Japan.

Crang, M., 2010. The death of great ships: Photography, politics, and waste in the global imaginary. *Environment and Planning A*, 42, 1084–102.

Crang, M. and Travlou, P. 2001. The city and topologies of memory. *Environment and Planning D: Society & Space*, 19, 161–77.

DeLyser, D. 1999. Authenticity on the ground: Engaging the past in a california ghost town. *Annals of the Association of American Geographers*, 89(4), 602–32.

Derrida, J. 1994. *Specters of Marx: The State of the Debt, the Work of Mourning and the New International*. London: Routledge.

DeSilvey, C. 2003. Cultivated histories in a Scottish allotment garden. *Cultural Geographies*, 10(4), 442–68.

DeSilvey, C. 2006. Observed decay: Telling stories with mutable things. *Journal of Material Culture*, 11(3), 318–38.

DeSilvey, C. 2007. Art and archive: Memory-work on a Montana homestead. *Journal of Historical Geography*, 33(4), 878–900.

Ecott, T. 2007. World's best wreck diving. The *Times*, London.

Edensor, T. 2005a. The ghosts of industrial ruins: Ordering and disordering memory in excessive space. *Environment and Planning D: Society & Space*, 23(6), 829–49.

Edensor, T. 2005b. *Industrial Ruins: Spaces, Aesthetics and Materiality*. Oxford: Berg.

Edensor, T. 2005c. Waste matter – the debris of industrial ruins and the disordering of the material world. *Journal of Material Culture*, 10(3), 311–32.

Edensor, T. 2007. Sensing the ruin. *The Senses and Society*, 2(2), 217–32.

Gordon, A., 1997. *Ghostly Matters: Haunting and the Sociological Imagination* Minneapolis: University of Minnesota Press.

Harrison, S., Pile, S. and Thrift, N. 2004. *Patterned Ground: Entanglements of Nature and Culture.* London: Reaktion.

Heidegger, M. 1962. *Being and Time*. Oxford: Blackwell.

Holloway, J. 2006. Enchanted spaces: The séance, affect, and geographies of religion. *Annals of the Association of American Geographers*, 96(1), 182–7.

Holloway, J. and Kneale, J. 2008. Locating haunting: A ghost-hunter's guide. *Cultural Geographies*, 15, 297–312.

Kean, J. 2009. *SS Thistlegorm*. Sharm el Sheikh: Farid Atiya

Lawton, L. and Weaver, D. 2001. Modified spaces. In *The Encyclopedia of Ecotourism*, edited by D, Weaver. Oxford: CABI, 315–25.

Lynch, K. 1972. *What Time Is This Place?* Massachusetts: MIT.

Maddern, J. and Adey, P. 2008. Editorial: Spectro-geographies. *Cultural Geographies*, 15(3), 291–4.

Merleau-Ponty, M. 1962. *Phenomenology of Perception*. Oxford: Routledge.

Peters, K. 2010. Future promises for contemporary social and cultural geographies of the sea. *Geography Compass*, 4(9), 1260–72.

Pile, S. 2005. *Real Cities: Modernity, Space and the Phantasmagorias of City Life*. London: Sage.

Pinder, D. 2001. Ghostly footsteps: Voices, memories and walks in the city. *Cultural Geographies*, 8(1), 1–19.

Stolk, P., Markwell, M. and Jenkins, J. 2007. Artificial reefs as recreational scuba diving resources: A critical review of research. *Journal of Sustainable Tourism*, 15(4), 331–50.

Vrana, K.J. and Halsey, J.R. 1992. Shipwreck allocation and management in Michigan: A review of theory and practice. *Historical Archaeology*, 26(4), 81–96.

Wylie, J. 2005. A single day's walking: Narrating self and landscape on the south west coast path. *Transactions of the Institute of British Geographers*, 30(2), 234–47.

Wylie, J. 2007. The spectral geographies of W.G. Sebald. *Cultural Geographies*, 14(2), 171–88.

PART III
Ocean Natures:
Mobilities and
More-Than-Human Concerns

Chapter 9

Sustaining Livelihoods: Mobility and Governance in the Senegalese Atlantic

Juliette Hallaire and Deirdre McKay

Introduction

Recent geographies of the sea emphasize the vital role that the ocean plays in the organization of societies. Steinberg argues 'the ocean is not merely a space used *by* society; it is one component of the space *of* society' (2001: 6 original emphasis). In this chapter, we explore the ways in which the sea is a space of society for Senegalese fishing communities, providing the resources essential to their everyday lives. We show how, in recent years, the depletion of marine resources, and competition for access to them, has created power struggles that are transforming traditional Senegalese uses and understandings of oceanic space through changed relationships with the state and increased mobility in order to sustain livelihoods.

Drawing on field observations and interviews conducted in Senegal, we show that for Senegalese fishermen the ocean was traditionally used as a resource space that enabled the development of local fishing knowledge and the projection of local political power. However, with increased resource scarcity the ocean has been reframed as a space of political conflict through power struggles and competition between various local actors and states. As such, fishermen begin to imagine the ocean as a transitional space organizing migration routes which enable them to earn the money necessary to reaffirm their control over their households, community and local fishing economy. In short, we examine how international maritime mobility is organized by small-scale Senegalese fishermen, describing fishing mobility from Senegal to Guinea-Bissau and Mauritania and maritime migration from Senegal to Europe. We show that these patterns of mobility are undoubtedly connected, but not simply reducible to fishers' resistance to state mismanagement and a lack of government attention to coastal communities (as Nyamnjoh, 2010, suggests). Rather, migration routes have become a central livelihood strategy that shapes land and seascapes for Senegalese fishers. Having experienced diminishing control over their households with declining fish catches, migration and its facilitation becomes the preferred strategy for fishers to retain, regain, or aspire to positions as household heads, providers and local political leaders. The sea, then, remains imagined as a space of possibility and potential that will solve the land-based problems of fishermen in the Senegalese Atlantic.

The chapter emphasises the entangled relationships between resource scarcity, the mobility of fishermen and governance of the sea. We first describe the way fishermen have developed mobility strategies to cope with the fishing crisis. Secondly, we examine sea governance and its effects, and how it has intensified with resource scarcity and the increased mobility described in order to both protect access to resources and/or to specific maritime areas. Finally, we show that large scale maritime migrations are empowering strategies that make border crossings valuable financial investments.

Coping with Resource Scarcity

In the case of Senegalese fishing, the sea is a space of political struggle. According to Deleuze and Guattari, the sea is a 'smooth space par excellence' (1988: 479) and provides the necessary fluidity for the deployment of nomads' movements. Like Senegalese fishers in the pre-colonial period, Deleuzian nomads follow resources without attending to the geography of state borders. When the state finally emerges to striate this space with an apparatus of political control, nomad-fishers respond by adjusting their 'lines of flight' – routes which stretch mobile actors between shifting origins and destinations (see Adey 2010). The different values given to fishers' mobility patterns by fishers themselves and by national governments have changed the geography of the Atlantic, and also changed relations between states and people, and the operation of states themselves. Though at first fishers' mobility was forced by depleting resources (signifying a lack of control over fish stocks), their mobility then became a way of resisting state power by organizing sea journeys to Europe. In this sense, fishers' mobility shapes the ocean space not just for themselves, but for their governments and those of European receiving nations. As Steinberg stresses,

> Because spaces (even the ocean) are simultaneously creations of social processes *and* arenas for everyday experiences, there is a constant negotiation between the 'spaces of representation' implied and reproduced by users of the sea (e.g. the world's navies, as well as its fishers, refugees, sailors, etc.) and the 'spatial practices' emanating from the structural imperatives of the world economy. (2001: 158)

Of a population of 12 million, more than 600,000 people in Senegal depend directly or indirectly on fishing for their livelihood (FAO 2010). The Senegalese fishing sector is organized according to a strict hierarchy, with power concentrated in boat owners, captains, and, finally, crewmembers on the basis of experience. Those who physically take part in activities on the water are from the last two categories. They generally work for boat owner who organize fishing activities from the land. Boat owners are former captains who are able to invest their income on fishing gear, canoes and motors. They hire, in order of priority, members of their immediate family and then more distant relatives belonging to the same community. Owners appropriate half of the value of the catch landed. The other half of the money is shared between the captain and crew. As soon as owners perceive a new constraint on their ability to see a return on their investment, they look for other investment opportunities. Owners are able to invest millions of CFA francs (FCFA) in maritime expeditions.[1] However, part of this enterprise is not controllable. Risks such as storms, illness at sea, death or bad fishing are taken into account in the organization of trips and are dealt through the spiritual work of the marabouts. Fishermen dedicate part of their investment to the consultation of a spiritual leader and gain control over the unpredictable through this investment.

Senegal's small-scale fishery sector currently counts around 18,916 *pirogues* (artisanal canoes).[2] This number of boats has tripled in the past 15 years, signalling towards the immense pressure on marine resources and the desperation of coastal communities. Although the largest canoes are 20 metres in length and motorized, their simplicity may makes them seem archaic to a Western eye. Yet fishers using them have sophisticated techniques and detailed local knowledge in which these boats are a key part of mobile resource extraction strategies. As highlighted by Chauveau (2000: 43):

> The organisational forms of small-scale fishery economy that accompany its growth are not 'modernised'; they remain traditional, and this is precisely for this reason that they are more efficient than the modern institutions and that they enable the development of the artisanal sector.[3]

There has, of course, been some engagement between the state and the small-scale, artisanal fishing sector. In the 1970s, the Senegalese government helped fishermen invest in their first motors, enabling them to go further out to sea to find fish stocks. From the 1980's, fishers organized the first motorized trips to foreign waters and started expeditions to waters off Guinea Bissau to catch high-value market demersal[4] species that had largely been fished-out in Senegalese waters. A number of fishermen abandoned the daily local fishing trips for these

---

1   The CFA franc is the currency used in Senegal and is guaranteed by the French Treasury with an exchange rate fixed to the Euro. In French it is abbreviated as FCFA.

2   Provisional results for the 2012 Senegalese Fisheries Registration Programme – statistics collected during Hallaire's interviews with fisheries officials, field notes, Dakar, 21 June 2012.

3   Translation by the authors.

4   Demersal species are found in deep waters. One of the most common in Senegal is the grouper (*thiof* in Wolof or *merou* in French) and this species is now highly threatened.

longer, riskier and more profitable expeditions throughout West African offshore waters (Failler and Binet 2010). This long-distance fishing migration has spread across Senegal, with fishers from Hann in Dakar, Saint-Louis, Mbour and Joal joining in. This form of regional mobility is comparatively new and distinct from the local everyday movement of small-scale fisheries in local Senegalese waters. Fishers organize journeys to what are now recognized as neighbouring countries' waters and then come back to Senegal to sell their high-market value catches. With fish lines, an average of 12 crewmen, and ice-boxes (to preserve the catch for the two weeks at sea), these boats would initially stay to fish for a couple of days before travelling back to Senegal to sell their catch. Now fishermen spend up to 15 days at sea using GPS to record their fishing places and find their way in the sea.

Regional mobility and new technologies have also seen the ways fishers map oceanic space change. Modou and Abdoulaye are two young fishermen who initially worked as crew members. They travelled by sea to Europe in 2006 and were deported back to Senegal two months later. When they drew a mental map of their local fishing area, Kayar, they explained how some of the ancient names given to these zones accorded to the land-based elements which could be seen in these fishing places. For example, one of the oldest Kayar fishing places is called 'Thiès', this because ancestors could see a big tree which they knew was located near the Senegalese city of Thiès. This type of naming provides information for navigation as, according to the position of the land-based elements spotted from the sea, fishermen can find their fishing place easily. Other names such as 'Mbayène', 'Palène' or 'Mbenguène' were those of influential families who fished in these particular places, whilst many names also refer to the fishing quality of the place: 'Takalé' means 'sparkling'[5] whereas 'Amul Yagal' literally means 'no patience' (as fish can be immediately caught there and thus fishermen need not show patience in these waters). However, these fishing places – and their traditional names – have diminished. New fishing places are a sign of the increased mobility of the fishermen. Their less poetic names (simply '11 kilometres' or '6 kilometres') refer to the distance that separates them from the shore and are indicative of the use of GPS devices in navigation.

In most communities, and especially in Kayar, fishermen have always combined their fishing activities with farming in order to complement their income (Le Roux and Noël 2007). However, increased population density has made access to land more difficult, which threatens the traditional balance between land-based and sea-based activities. Fishers often do not want to disclose their new fishing places so that they can keep the preserve of their exploitation. In this case, they give them new 'personalized' names. These practices lead to a new individualization of the fishing activity and to a complex geography of the sea space as its knowledge is less shared. Names *and* new fishing places now belong to individuals or small groups of fishermen who do not wish to share their new

---

5   This translation can be discussed as 'takale' can also mean 'to gather'.

information. Therefore, particular power/knowledge relationships emerge from this ecological crisis. Knowledge and power are produced through the increased mobility of the fishermen: controlling knowledge such as information about fishing places is, for fishermen, a way to secure resources and maintain a balance in terms of the exploitation of the sea.

Mobility and Sea Governance

The regional fishing strategies previously described, were first developed by a small number of fishermen, however, the numbers involved expanded when, in 1989, a border conflict broke out between Senegal and Mauritania. In 1991, Mauritania closed its maritime border and stopped allocating free fishing licences to Senegalese fishermen who traditionally fished in Mauritanian territorial waters (Diop 2004). Tensions between Senegalese and Mauritanian boats and between the national governments finally led a number of Senegalese boats directing their fishing trips further south to waters off Guinea Bissau. Although not all Senegalese fishers are long-distance fishers, those following such strategies are now found in almost every Senegalese port. Indeed, Senegalese fishermen always come back to Senegal and choose the best place to land their catches to sell into the national market.[6]

Fishermen who now travel to Bissau-Guinea need high levels of investment. In the Petite Cote, a local boat owner told how he controls an informal company of around 110 fishermen from his home village. Each of his nine canoes requires around FCFA 600,000 for its petrol, and more money for the kilos of rice, water, sugar and tea that provision its crew for the time spent at sea. Each canoe also needs a FCFA 900,000 licence from the Guinean government to enable crews to fish legally in Guinean waters. This example demonstrates the governmental regulations fishers must navigate. This particular owner explained how he sought to gain his fishing licenses without the help of the Senegalese Ministry of Fisheries.[7] In 2011, some Senegalese boat owners had been sold invalid fishing licences and had been forced to pay fines to the Guinean authorities. A group of affected owners had asked the Senegalese government to help them by accessing valid licences for 2012, directly from the Guinean authorities. However, this boat owner had his own reliable broker to obtain licences from the Guinean government faster than through the Senegalese administration.

These changes to fishing and the threats it places upon sustaining livelihoods, has led to internal migration in Senegal. Following the 1989 conflict and the resulting

---

6   However, a key informant reported that around 11 ice-box canoes currently fish illegally in Mauritania, navigating 200 km to the west and 500 km to the north in order to avoid Mauritanian border controls. These long journeys involve huge amounts of petrol, ice for the fish, and supplies for the 10 day trips for the 12 crew members.

7   Field interview, April 2012

closure of Mauritanian maritime border, a huge number of Guet Ndarians[8] decided to organize fishing trips to Guinea Bissau but leaving from the ports of Saint-Louis or from Dakar – two areas much closer to Senegal's southern border.[9] Thus, after 1989, Guet Ndarian fishermen have spread down the Senegalese coastline, relocating their households to new ports as internal migrants. Because they were landing catches from Guinea Bissau, rather than competing with local fishers in their new settlements, Guet Ndarian migrants were mostly tolerated. Although they are respected and considered to be 'the best fishermen' of the region, they are also perceived as very rough, stubborn, and unpredictable, so their integration in new settlements was not always successful, even if they were landing high market value fish from long distance trips and were not in competition with local fishers. Thus the Guet Ndarians have become a highly mobile group, both along coasts and the open ocean off West Africa.

Those fishers who remained settled in Saint-Louis have developed other fishing strategies. Purse seine[10] fishermen buy official fishing licences for pelagic[11] species or go fishing illegally in Mauritanian waters when they do not have licences. Line fishermen cross the border illegally as Mauritania does not provide licences for demersal species. They then travel further north and west than before the 1989 closure to avoid being discovered by Mauritanian state patrols in the night. The other strategy they pursue is to work for Mauritanian 'businessmen' who have legal Mauritanian licences for line fishing. Their business strategy is to hire Senegalese fishermen to fish each day in Mauritanian waters and the fishermen then have to live in camps not far away from Nouakchott in Mauritania. Fishers following this strategy stay in Mauritania with their crew and canoes for one or two months before coming back to Saint Louis on a regular basis (Binet et al. 2010 and field interviews).

These accounts highlight how state regulation can be easily escaped. Yet the state is nonetheless present at sea. The Senegalese government has intensified competition for fish stocks by enabling foreign fleets to fish in Senegalese waters, signing a growing number of industrial fishing agreements (Catanzano and Rey Valette 2002). The government has also attempted, at the same time, to regulate local-scale fisheries by controlling access to the sea, fishing techniques and fishing areas. Since access to the sea is traditionally open to everybody in Senegal, fishermen find it hard to accept government restrictions. The ambiguity of Senegalese maritime governance has generated frustration among the fishermen

---

8   Fishers described as 'Guet Ndarian' come from Guet Ndar, a famous fishing village of the spit of land stuck between the Mauritanian border and the mouth of Senegal River. This village has high population density, a long fishing history and occupies a very narrow area, leading to intense resource competition. These factors have forced the fishermen from St Louis to expand their fishing grounds and migration routes. Fishers from St Louis are well-known across West Africa for their great navigation and fishing skills.

9   Field interviews, Saint-Louis, Dakar, 2011 and 2012.

10   Large fish nets used to catch species close to the surface.

11   Fish species living near the surface.

who have subsequently organized themselves through a number of national professional associations. These organizations lobby the government, calling for more coherence, transparency and government attention for the sector.

The growing frustration of the fishermen slows the efficient application of fishing regulation. However, from the viewpoint of government agents, small-scale fisheries' recent evolution is threatening the efforts of government to managing the resource. According to an official in charge of Senegalese maritime fisheries, it is the small-scale fishery which causes conflicts occurring with the international trawlers:

> For example, the nets: they generally let their nets without indication. Then, when the industrial trawlers pass, they can't see the nets, they are not visible to the naked eye, so they tear the nets up, which then causes all kinds of conflicts.[12]

Because fishers are now so mobile, they are more and more difficult for the government to regulate. Yet these regulations undermine government claims to understand the sector. Among the new fishing rules, for instance, is one specifying fishermen now must wear lifejackets. However, they often refuse to do so considering that this questions their ability to navigate safely and would moreover, interfere with God's will (Sall 2007). They also claim lifejackets slow down work on board.

Fishers now cross international boundaries, escaping surveillance and navigating for more than 15 days thanks to the accurate use of GPS combined with a developed sense of stellar navigation and a particular maritime culture (Sall 2007). Fishers can quickly adapt their movements to constraints and lack of resources in local Senegalese waters on the basis of rational calculations. Their small wooden canoes enable them to travel faster and at a lower cost than modern boats with heavy, expensive gear. Over time, these fishers have gained in-depth knowledge and navigation skills that are now essential for their organization of illegal migration to Europe.

Empowering Maritime Migration Strategies: When Crossing Borders becomes Synonymous with Valuable Investments.

From the late 1990s, fishers identified carrying migrants as a potential strategy to work around the constraints of diminishing catches, the challenges in obtaining licenses, and intensified fishing regulation, and compensate for the decline in their income from fishing. Thus the organizers of the first migration voyages were either boat owners or comparatively wealthy fishermen with funds to invest. The organizers would stay in Senegal, hiring others to take the risks of the voyage. Usually they invested in 20 metre boats, motors, supplies for the passengers, and hired skilled captains willing to go to Europe. Those best qualified as captains were fishers already familiar with the long-distance fishing routes up the West

---

12   Field interview, Official of the *Direction des Pêches Maritime*, Dakar, June 2011.

African coast and accustomed to evading maritime surveillance. Would-be migrants willing to embark on these voyages were mostly young to middle-aged Senegalese men – fishers involved in local and regional scale fishery or non-fishers who could not sustain their livelihoods. Many of the fishers who became would-be migrants had been working as daily-paid crew on purse seine canoes or on small line-fishing canoes and, as stocks declined, found their labour was no longer profitable or needed. Determined to go to Europe, an average of 100 potential migrants paid FCFA 400,000 (€600) in passage on each boat. Samba was a fisherman who left Senegal with the first migrant-carrying canoes. He wanted to work in ports in Europe, sending his earnings to his family before returning to Senegal. He explained that he was not afraid of crossing the ocean: 'we know the sea, sometimes we spend 15 days at sea to fish, it doesn't scare us.'[13]

Maritime migration from Senegal to Europe further intensified after land-based routes through North Africa were restricted. For example, a number of would-be immigrants to Europe from sub-Saharan Africa crossed the border into Spanish territories on the African continent breaking through the boundary fences surrounding the Spanish exclaves of Ceuta and Melilla in Morocco. Reportedly, their assault on the fences was brutally prevented by Spanish and Moroccan authorities (see Carling 2007, De Haas 2007). The failure of this attempt to breach the land border led to the organization of new maritime routes to Europe, targeting the much more permeable ocean border, departing first from Mauritania and then from Senegal. At the same time, in 2004–2006, the use of GPS devices was spreading among Senegalese fishermen communities, their navigation skills rapidly developing the capacity to expand their long-distance mobility. In addition, the stories told by supposedly successful returned Senegalese emigrants from previous rounds of migration to Europe convinced younger generations that migration was a route to success. Even imagining a personal migration project to Europe became considered a quick and easy way to gain respect in the community. Those migrants who were sending remittances home on a regular basis had already secured themselves a respected position, even if they had been absent for several decades (Fouquet 2008). The combination of politics, technology, economics and social conditions encouraged fishers to take part in the organization of migration journeys to the Canary Islands.

Cheikh, a returned migrant-fisherman from Dakar, explains that he took part in an illegal journey to the Canary Islands. His boat had a problem and the crew decided to land in Mauritania without even having reached the archipelago. His testimony shows how this decision involves his whole family and becomes a potential solution for a better life:

> It's been 25 years that I have been going to the sea and I've got nothing. We don't have means. Nobody can help me, my family believes in me; they depend on me. If I go to Europe, what I will earn, I will send it. At the moment, the sea provides me with 3000 Francs a day, sometimes nothing. I come back, I have

---

13   Field interviews, Dakar, 2007

nothing, my family waits and I still have nothing. I meet with my friends, they help me a bit but I am ashamed. What is good for me is to fight to earn a living.'[14]

In 2006, Modou crossed the ocean with other migrants in order to reach Europe. A 20 metres long fishing boat driven by experienced captain fishermen led him to the Spanish archipelago. In charge of his family since he was able to earn a living, he faced the decrease of fish resources and was no longer able to provide a decent livelihood to support his relatives. He decided to embark for this perilous illegal journey and fund it with the sale of his fishing gear. Once in the Canary Islands after seven days spent in high sea, he got arrested and was sent back by the Spanish authorities to Kayar, his village. His perception of the decline of resources is fairly accurate as he assesses it with the approximate length of his fishing trips, the number of fishing places he has to go to and the strong feeling that the sea is no longer profitable for him. As most of the returned fishermen who were interviewed, the difficulties Modou is facing at sea and the low average income he now earns, were the main reasons for his migration project.

Crossing the ocean becomes synonymous with a positive fight, an opportunity to recover the role that the sea is no longer able to provide to him. Although the sea, as a resource place is no longer fulfilling its function, it remains a central 'socially constructed' space enabling the fishermen to 'become someone' (this expression was recurrent in the interviews). This migration project takes the meaning of a 'sacrifice' which young men have to make for their family as simply put by Pape, a repatriated fisherman from Dakar suburbs (Yoff): 'there is nothing more in the sea and there isn't a lot of land for the children. There is no land in Yoff, we are obliged to go sacrifice ourselves.' His friend adds 'It's a matter of dignity'.[15]

As a response to such migration practices, border controls have been reinforced with the development of Frontex (the European border agency) in Spanish, Mauritanian and Senegalese waters. As such, migratory departures were organized further south in Senegal, from Kayar, Dakar, Mbour and Ziguinchor by fishermen. The further south the voyages departed, the higher the risks however. These departures were constantly adjusted to maritime border controls and boats would often leave secretly at night trying to reach international waters undetected to escape increasing numbers of patrols. Moreover, through the constant adjustment of the routes to border controls, fishermen were resisting maritime controls as well. In response, European border agency adopted a similar mobility strategy to arrest those undertaking illegal migration journeys. The Spanish government counted that 90,000 illegal migrants had reached the archipelago between 2001 and 2010.[16] Among them, thousands were Senegalese and were deported back to Senegal through the frame of Senegalese/Spanish cooperation.

---

14   Field Interview, Yoff, Dakar, 2007.
15   Field Interview, Yoff, Dakar, 2007.
16   'Lucha contra la Inmigración Ilegal, Balance 2010', Ministerio del Interior, Spain, www.mir.es, reporting official Spanish government statistics.

Senegalese people very commonly called the phenomenon of illegal migration 'Barça or Barsakh' meaning 'Barcelona or the Beyond' (see Bouilly 2008). This appellation associates the theological and cultural significances of the journey and converts the ocean either into a pathway to Europe – better nicknamed as 'Barça', in reference to the football club of Barcelona- or, more tragically, to the Beyond (in Arabic). This expression also shows that migrants were aware of the high risks they were taking. Although they put their life at great risk and had a high probability of being repatriated, they relied on God's will and were still convinced they had to leave. Most of the migrants had consulted a marabout before taking the decision to embark in these boats. They trusted their spiritual leader who was able to let them know whether they should take this risk and therefore get particular protections. Marabouts also provided *talismans* which would make them invisible to police patrols.[17] In the case of an accident in high sea, their death would often remain ignored, leaving desperate families with the hope of an everyday more unlikely return.

Conclusion: Productions of Knowledge, Projections of Power

In the case of the Senegalese fishing community, the ocean's function has unquestionably changed over time. The only constant is that the ocean remains the decisive space embodying power over life and death, where determined fishermen struggle for the future of Senegal's coastal communities.

On a European level, the deployment of Frontex changes the ocean into an external space of security on which border practices protect the European territory. The 'lines of flight' adapted by fishermen to avoid border controls have, in response, seen Europe produce a further 'striating' action on the formerly smooth oceanic space (Deleuze and Guattari 1988), intensifying surveillance on the open ocean, monitoring departures and arrivals up and down the Atlantic coast, and scaling-up inspections at ports. By constantly improvising and adjusting their routes, fishers engage in power struggles with the border agents charged with blocking their trajectories. Yet these blockades make the promises of boat owners and voyage organizers valuable, as stories of successful border crossings, mean that young fishers may make two or three attempts at the voyage.

For the subregional West African governments – Senegal, Mauritania, and Guinea Bissau – the sea is both a space to protect and a source of revenues. By controlling the growing mobility of fishermen, their use of gear, their safety equipment and requiring them to hold permits to fish, fisheries ministries and their officials seek to preserve a disappearing natural resource. At the same time, other government agencies seek to earn revenue by contracting fish resources to foreign boats through joint ventures. Governments are increasingly able to access funds for migration prevention, resettlement and local development from European sources, in return for better governance of fishers and local policing of migration departures, both on the coasts and at sea. So the ocean space off West Africa

---

17   Field interviews, Dakar 2007, 2011; see also Naranjo Noble 2009.

becomes doubly striated through the governance of mobility, both for fishing and migration at sub-regional and European level.

But these comparatively new projects of governance may be somewhat ambivalent. National officials who now seek to extract fines from highly-mobile fishers who lack the proper permits and from foreign governments via migration-prevention funds are well aware that both these streams of value depend on migration to Europe continuing to be the preferred solution to the woes of Senegal's coastal communities. Many Senegalese – not just fishers – have invested in the shared imaginary of an ocean crossing as the best route to a more secure livelihood, increased social status and thus a better life.

The social construction of fishers' imagined maritime space here shows how the ocean takes on the characteristics of a Foucauldian heterotopia (Foucault 1994, Steinberg 2001), the maritime space acting as a mirror of the land. Foucault (1994) uses the metaphor of the mirror to illustrate how heterotopias operate. The heterotopia/mirror reflects the dematerialized image of a space and, through this reflection, the identity of the initial space is recreated but simultaneously subverted. Here, the organization of maritime space mirrors then transforms that of the land. Indeed, we have shown how Senegalese fishers organize ocean space in reference to their terrestrial territories and landed commitments. Foucault argues that heterotopias are at the core of social practices and there is a continuous movement between them and the central places of societies, with heterotopias playing a decisive role in the creation of social identities.

In coastal Senegal, we have shown how sea space remains central in fishing communities, whatever role and function it embodies. Crises, conflicts, technological and geographical changes and governmental regulations occurring in oceanic spaces have had meaningful consequences in the social organization on and across landed spaces. As such, as soon as the security of the maritime space is jeopardized, as we have seen in the Senegalese fishing crisis, power relations on land are reconfigured. Such shifts then give a new shape to the seascape, in this case moving towards potential migratory routes to a better life.

References

Adey, P. 2010. *Mobility*. London: Routledge.

Binet, T. et al. 2010. *Evolution des Migrations de Pêcheurs Artisanaux en Afrique de l'Ouest depuis la Fin des Années 1980*, IUCN / CSRP. Rapport n° 2 de l'étude relative à l'état des lieux et l'évolution récente des migrations de pêcheurs artisans dans les pays de la CSRP. Dakar.

Bouilly, E. 2008. La Couverture Médiatique du Collectif pour la Lutte contre l'Immigration Clandestine de Thiaroye-sur-Mer (Dakar- Sénégal). Une Mise en Abîme du Discours Produit au «Nord» sur le «Sud». *REVUE Asylon(s)*, 3. Available at: http://www.reseau-terra.eu/article721.html [accessed: 26 January 2012].

Carling, J. 2007. Unauthorized migration from Africa to Spain. *International Migration*, 45(4), 3–37.

Catanzano, J. and Rey Valette, H. 2002. Histoire des Pêches, Accords de Pêche, & Politiques Publiques en Afrique de l'Ouest. In *Proceedings of the International Symposium of Dakar – 24–28 June*. Dakar, 475–94

Chauveau, J.-P. 2000. *Les Pêches Piroguières en Afrique de l'Ouest. Dynamiques Institutionnelles: Pouvoirs, Mobilité, Marchés*. Dakar: KARTHALA Editions.

Deleuze, G. and Guattari, F. 1988. *A Thousand Plateaus: Capitalism and Schizophrenia*. London: Athlone Press.

Diop, O. 2004. Migrations et Conflits de Pêche le Long du Littoral Sénégalo-Mauritanien: le Cas des Pêcheurs de GuetNdar de Saint-Louis (Sénégal). *Recherches Africaines*, 3, 1–16.

FAO. 2010. *The State of World Fisheries and Aquaculture*, Rome: FAO Fisheries and Aquaculture Department. Available at: http://www.fao.org/docrep/013/i1820e/i1820e00.htm [accessed: 26 January 2012].

Failler, P. and Binet, T. 2010. Sénégal: Les Pêcheurs Migrants: Réfugiés Climatiques et Ecologiques. *Hommes et Migrations*, 1284, 98–111.

Foucault, M. 1994. Des Espaces Autres. In M. Foucault, ed. *Dits et écrits (1984)*. Paris: Gallimard, Nrf, 752–62.

Fouquet, T. 2008. Migrations et 'Glocalisation' Dakaroises. In *Le Sénégal des Migrations: Mobilités, Identités et Sociétés*, edited by M.C. Diop. Dakar: KARTHALA Editions, 434.

De Haas, H. 2007. *The Myth of Invasion: Irregular Migration from West Africa to the Maghreb and the European Union*, International Migration Institute: University of Oxford.

Mercier, P. and Balandier, G. 1952. *Les Pêcheurs Lebou du Sénégal: Particularisme et Evolution* Études Sén., Saint-Louis, Sénégal: Centre IFAN-Sénégal.

Naranjo Noble, J. 2009. *Los Invisibles de Kolda. Historias Olvidadas de la Inmigración Clandestina.* Atalaya, Barcelona: Peninsula.

Nyamnjoh, H.M. 2010. *'We Get Nothing from Fishing': Fishing for Boat Opportunities Amongst Senegalese Fisher Migrants*. Bamenda, Cameroon: Langaa Research and Pub. Common Initiative Group.

Le Roux, S. and Noël, J. 2007. Mondialisation et Conflits Autour des Ressources Halieutiques. *Ecologie & Politique*, 34, 69–82.

Sall, A. 2007. Loss of bio-diversity: Rrepresentation and valuation processes of fishing communities. *Social Science Information*, 46(1), 153–87. Available at: http://ssi.sagepub.com/cgi/doi/10.1177/0539018407073663 [accessed: 30 November 2011].

Steinberg, P. 2001. *The Social Construction of the Ocean*. Cambridge: Cambridge University Press.

Chapter 10
Governance of the Seas: A More-Than-Human Perspective on the Cardigan Bay Scallop Fishery

Christopher Bear

Introduction

So far in this volume, we have heard the voices of human actors and their engagement with the sea (see for example the last section). Our discussion of water worlds, then, are almost always human centred. But what of other actors at sea, the non-humans that often dwell below the surface? This chapter shifts the focus to the more-than-human world of the sea, focusing in particular on the role of non-humans in practices of fisheries management. Sea fisheries management has been increasingly characterized by a shift from government – the regulation of fisheries by sovereign states and international governments – to governance, whereby an increasing heterogeneity of state and non-state actors work together in the development and implementation of regulations. The constitution of such governance structures takes many forms, with non-governmental and for-profit institutions playing roles of varying importance. In spite of this increasingly – or ostensibly – inclusive approach, fisheries regulation and the debate that surrounds it remain, in many ways, relentlessly anthropocentric.

Given the central role of fisheries biology in the management of sea fish stocks, this argument might initially appear somewhat spurious. However, viewed in the light of ongoing debates around the 'more-than-human' (Whatmore 2006), it might be argued that focus often lands on the rights of fishermen and the economic contribution of fisheries rather than on either the rights of fish or, indeed, their contribution to the debate (see also Jones 2000).

In this chapter I engage with Latour's ideas around the recognition of non-humans as actors in a redefined 'citizenry' (2004: 62), considering some of the ways in which fish (and other non-humans) have been (un)wittingly silenced in much discussion. Through this chapter, I show some of the ways in which non-humans 'contribute [their] share' (Latour 2004: 58) through their actions and interactions. The chapter, therefore, moves away from existing work on the geographies of the sea, which has tended to focus on it as a space to be crossed or as a space which is socially constructed (Steinberg 2001). While considering the importance of such ideas in debates and practices of management and regulation, the chapter foregrounds the 'liveliness' (Lorimer 2010; see also Peters 2010) of the sea, examining the disruptive and mobile tendencies that characterize the space and its inhabitants. I use examples of disruption by, and mobility of, non-humans to argue that anthropocentric approaches to the analysis of the sea are partial and misleading. I further argue that a revitalized more-than-human analysis of sea fisheries also has implications for their management and, especially, for how future debates around and practices of governance might proceed.

To illustrate and develop these arguments, the chapter draws on the case study of the Cardigan Bay scallop fishery in mid-Wales. There has been considerable controversy around the management of this fishery, largely as a result of interactions between scallop dredgers, scallops, bottlenose dolphins and the seabed. Following a brief literature review that outlines key features of recent debates around fisheries governance, as well as discussing the implications of 'more-than-human geographies', the Cardigan Bay fishery is examined through three modes of analysis. The first adopts a descriptive and anthropocentric approach, introducing the key institutions involved in the management of the fishery and outlining some of the central debates around this. Subsequent sections demonstrate that such approaches are not sufficient in understanding how the debate has played out. In these sections, the active role of non-humans is foregrounded, initially through a focus on disruption and dissidence and subsequently by looking at the role of movement. In the conclusion, I discuss the implications of this more-than-human approach for the analysis and practice of future fisheries governance.

Governing Sea Fisheries

Environmental issues are increasingly dealt with through practices of governance, providing 'a third way between the two poles of market and state, incorporating both into a broader process of steering in order to achieve common goals' (Evans 2011: 4). As such, the practice of governing increasingly involves a range of non-state actors, such as 'charities, NGOs, businesses, and the public' (ibid), often working in partnership or alongside the state, at varying scales from the local to the global. Such movements characterize many current fisheries management regimes, where 'institutions and individuals that are not harvest rights holders, or mandated centralised management agencies' now have an 'increasing role and influence' in many fisheries (Gibbs 2008: 115–16). For Gibbs, the actors now include 'NGOs … local informal institutions, collectives of processors and support industries, the general voting community, local residents in fishing communities, recreational fishers, and global communities who influence fisheries' (ibid). Such links can be identified, for instance, in the 2012 reform of the EU's Common Fisheries Policy, whereby celebrity-led campaigns, attracting mass public support, have entered discussions alongside more established concerns of fishing communities, scientists and politicians, effecting change especially with regard to discards of fish at sea (see, for instance, Meade 2012). Management decisions, in other words, are increasingly not top-down but 'are emergent properties of the network' (Gibbs 2008: 116).

Geographers have approached the governance of the sea from a range of perspectives, although it should be noted that their engagement with this topic remains relatively limited. Some of the most prominent work has focused particularly on neoliberal approaches. For instance, in an important series of papers, Mansfield (2004a, 2004b, 2007a, 2007b) has examined the changing nature of property rights in ocean fisheries, showing how the understanding of fisheries as common property resources was increasingly challenged by fisheries economists in the 1950–1960s as 'market failure' (2004a: 316). This common property regime was widely replaced from the 1950s to the 1980s by an increasing focus on state control, where 'individual states extended their political economic jurisdiction generally from three nautical miles from shore to 200 nautical miles', with these latter limits becoming 'customary international law in the early 1980s' (Mansfield 2004a: 317). Mansfield views this expansion partly as a contradiction of neoliberal approaches, through its extension of State powers, but equally argues that this enclosure of global commons 'provides the foundation from which states can further enclose the oceans through limited licences or other privatisation schemes' (ibid). Through this work, Mansfield has shown how the role of the state in ocean fisheries management is increasingly contingent, with management practices increasingly developed by, or with, private fishing companies.

The role of the market, and a more dispersed version of network governance, has been emphasized in a second body of literature, which has examined product labelling schemes, such as that operated by the Marine Stewardship Council (MSC) (e.g. Constance and Bonnano 2000, Eden and Bear 2010, and Hatanaka, Bain and Busch 2005). Through the labelling of produce from certified fisheries, this scheme aims to make it easier for consumers to identify fish that has been harvested sustainably. By doing so, it is designed to pressure other fisheries into changing their practices to achieve certification. As such, it harnesses the power of the market to directly impact on fishing practices. In doing so, it not only affects fishery management institutionally, but also brings consumers, if indirectly, into a governance framework (Gibbs 2008: 117).

A third area of literature has focused less on the market and more on the role of fishing communities, questioning how these might be more actively brought into management discussions (St Martin 2006, St Martin and Hall-Arber 2008). Such literature has tended to focus on the notion of 'co-management', viewed initially as 'the strengthening of the local perspective, in the sense that the community should be restored as a key domain in management' (de Vivero et al. 2008: 322) but increasingly extended to 'all the social and institutional agents that have a legitimate right to be consulted in fisheries policy decisions' (ibid).

The fisheries management literature, therefore, is increasingly characterized by the inclusion of heterogeneity of actors, whether institutions or individuals, state or non-state, directly or indirectly. In spite of a move away from truly 'common property' approaches, the range of actors involved in the management has grown, rather than shrunk. However, in spite of this apparent inclusivity, the voices of the creatures being managed – the fish in particular – often seem not to be heard, other than through the various representations produced by the different interests. In the next section, I introduce an area of the geographical literature that has attempted to problematize this situation.

More-Than-Human Geographies

A long-standing interest of human geographers is the relationship between humans and 'nature'. This relationship has been explored through a variety of theoretical lenses, including social construction and Marxism (see Castree and Braun 2001). The work has increasingly been characterized by a desire to tease apart widely-accepted dualisms, such as society/nature, humans/environment. As such, authors have followed Latour's (1993) call to view such categories as convenient labels, associated with modernity, and instead to investigate the processes that have led to their use and acceptance. A key strand of research within this literature has focused on animals. This initially stemmed from a concern that lumping rocks, trees, animals, water and suchlike together under the meta-category 'nature' was unhelpful (Philo 1998) and did not reflect the everyday interactions that take place (Philo and Wilbert 2000); in many ways, it is a rather abstract term.

In such a way, this work set out to 'bring the animals back in' to 'human' geography (Wolch and Emel 1995), re-constituting 'society' as 'more-than-human' (Whatmore 2006; see also Latour 2005). Some of this literature on 'the new animal geography' (Philo 1995) has looked at different representations of animals (e.g. Gullo et al. 1998), but much recent work has turned to consider non-human actions, the coproduction of place and practice by humans and non-humans (Matless et al. 2005, Lorimer 2006, Lorimer 2008, Bear and Eden 2011), and the form and disciplining of non-human subjectivity (Holloway 2007). Much of this research has highlighted the active participation of non-humans in the heterogeneously reconstituted society, where they are not merely passive surfaces onto which humans paint their desires and beliefs (Philo and Wilbert 2000), but agents that disrupt, transform and create, and which are beings in their own right. As such, some writers have attempted to develop more-than-representational (Lorimer 2005) methodologies for studies of the more-than-human; Hinchliffe et al. (2005), for instance, in their study of water voles, combined participant observation and interviews with the 'reading' of water vole 'writing' (their spraints and footprints). Others have adopted a 'critical anthropomorphism' (Johnston 2008), engaging with the experience of humans who spend considerable time in direct contact with animals (see also Bear 2011, Bear and Eden 2011, and Lorimer 2006,). This engagement with animal practice has led to further problematization of convenient labels, with various authors (e.g. Lulka 2009, Philo 2005) calling for greater emphasis on non-human difference.

While increasingly methodologically innovative and ontologically insightful in developing a more symmetrical approach to human-nonhuman relations, fewer studies have engaged directly with the practical implications of these shifts in relation to the politics of environmental management. This is not to say that such issues have been ignored. Wolch (2002), for instance, suggested the need for a 'new urban democracy' that would, in some way, include non-humans alongside humans. She was frustratingly vague about what form this new democracy might take, however. Her argument, in other words, was more ontological than practical. Hinchliffe took such arguments forward, considering 'what shape a politics involving non-humans might start to take' (2008: 89), demonstrating some of the ways in which 'natures are in the making' (2008: 88), as opposed to pre-set and given. As such, he engaged with Latour's contention that the 'modernist "Constitution"' (2004: 54) might be replaced through 'the reunification of things and people, objects and subjects' (2004: 57). Latour's rationale is that the modernist Constitution involves 'a ruinous anthropomorphism through which objects, indifferent to the fate of humans, were in the habit of intervening from the outside and acting without due process to sweeping away the work of political assemblies' (2004: 54). Latour, in other words, argues that practices of science, conservation and environmental management have tended to essentialize and stabilize non-humans as objects to be studied, controlled and managed, and their actions, interactions, emotions and subjectivities have, in many ways, been silenced:

> the assembly of humans is finding itself obliged to reconsider the initial division, and it is asking the other assembly, which has been meeting in secret for centuries and whose political work has always been hidden up to now, to contribute its share. (2004: 58)

In *Politics of Nature*, he explores the implications of such contentions, reflecting on how politics might be extended to include humans and non-humans alongside each other. His approach to the problem is to recognize 'the citizenry' by defining 'the collective as an assembly of beings *capable of speaking*' (2004: 62 original emphasis). Here, speech is not merely a vocal articulation but should be understood relationally: 'no beings, not even humans, speak on their own, but always *through something or someone else*' (2004: 68 original emphasis). Communication, in other words, is not through vocalization alone. Hinchliffe (2010: 306) summarizes the implication of such an argument: '... politics is about more than words and more than representation – it can also be about how things are done in ways that could be otherwise and about struggles between different enactments of reality'.

The remainder of the chapter explores this alternative conception of 'society' more thoroughly through the specific example of the Cardigan Bay scallop fishery, attending to the redefined more-than-vocal and more-than-representational politics. Through this, I show how fisheries management is often discussed in anthropocentric terms before exploring some of the ways in which fisheries might be approached more symmetrically.

The Cardigan Bay Scallop Fishery

The Cardigan Bay scallop fishery is a good illustration of the ideas outlined in the previous sections. In many ways, it embodies the shift from common property to state regulation to network governance, as seen in many other fisheries. The number and diversity of non-state actors that contribute to changing management practices have peaked with the debate that has surrounded practices of dredging. In this section, I introduce the fishery, along with the recent controversies, through a political economic approach that foregrounds institutions and, in particular, fishermen and fishing communities. Subsequent sections move on to show how inadequate and two-dimensional such an approach is, before finishing with a discussion of the implications.

Cardigan Bay stretches approximately 60 miles from Bardsey Island (in north Wales) to Strumble Head (in south Wales) and is an inlet of the Irish Sea. The Bay is home to a wide range of inhabitants and activities, which has led to a complex management structure encompassing a wide range of institutions. The Bay is increasingly marketed to tourists for a range of water sports, but maintains a significant fishing industry, particularly focused on scallops. Alongside this, the Bay is renowned for its population of over 300 bottlenose dolphins – one of only two concentrated populations in the UK (Cardigan Bay SAC, date unknown) – which also draws tourists to the area. There is clear potential for conflict here between these different uses and users, but it is the relationships between scallop dredging and dolphin habitat that have proved especially controversial.

The ecological significance of the dolphins and their habitat received official recognition in 2004, with the designation of part of Cardigan Bay as a Special Area of Conservation (SAC) (Ceredigion Council et al. 2008). These Areas are established through the EC's Habitats Directive and have the intention of maintaining or restoring 'habitats and species of European importance to a Favourable Conservation Status' (ibid: 8). The SAC was designated especially to protect the population of bottlenose dolphins, although other species – Atlantic grey seals, river lamprey and sea lamprey – along with reef, submerged sandbank and sea lamprey habitats were also included (ibid). The boundary of the SAC forms a rectangle, extending from Aberarth in the north to Ceibwr Bay in the south, and 12 miles out to sea (ibid: 18).

Scallop fishing in the Bay has undergone dramatic changes over the past three decades. In global, or even national, terms, the Cardigan Bay fishery was of negligible economic value until 1980, at which time 'dense beds of scallops' were discovered (Anon 2010: 5). Seventy fishing vessels descended on the Bay from around the UK, landing over 1,000 tonnes of scallops, drastically reducing the stock and limiting catches in future years. By 2000, the catches had risen again to 300 tonnes but increased abruptly and markedly in 2008, when 3,500 tonnes were landed. By 2008, the Cardigan Bay fishery represented 30% of the total scallop catch for England and Wales (ibid). Although the intensity of fishing effort has increased, its spatial extent remains relatively stable, largely the result of the relatively sedentary nature of adult scallops.

Scallop fishing in the Bay is carried out under the framework of the EU Common Fisheries Policy but within 12 nautical miles of the coast (i.e. Welsh territorial waters) is regulated directly by the Welsh Assembly Government (WAG). Between 2005 and 2010, this regulation was through the Scallop Fishing (Wales) Order 2005, which instituted a close season for fishing between 1 June and 31 October (based roughly on the scallops' reproductive cycles), introduced restrictions on fishing technologies (for instance, preventing boats from towing more than 16 scallop dredges and restricting the size of dredges that could be used) and prohibited the carrying of scallops with a diameter of less than 110mm. The Order was introduced both to protect the scallops and to tailor the regulation of the fishery to Welsh requirements. Until April 2010, much of the regulation of fishing practice in the Bay was devolved to the North Western and North Wales Sea Fisheries Committee, which was composed of councillors from coastal counties, representatives of the Environment Agency and Members of WAG (North Western Inshore Fisheries and Conservation Authority, date unknown). This was disbanded in April 2010, following the repeal of the Sea Fisheries Regulation Act 1966 in Wales (The North Western and North Wales Sea Fisheries District [Consequential and Transitional Provisions] Order 2010), at which point WAG itself took over regulation of sea fisheries around Wales.

The debate about scallop dredging intensified towards the end of the decade, with various groups mounting prominent campaigns for and against the fishery. The 2005 Order had been introduced 'to limit effort and secure a future fishery' but WAG (2009a) felt that this 'did not go far enough' as 'efficiencies in fishing methods and practices' (i.e. dredging) increased. Concerns around the fishing were both socio-economic and environmental. With regard to the former, there was particular controversy around the increasing number of fishing vessels in the Bay from outside Wales. Across Wales as a whole, the number of vessels fishing for scallops registered outside the country rose from 8 in 2005 to more than 60 in 2008), and the local press reported in 2009 that 40 boats had been displaced from a scallop fishery in Devon which had recently been closed for conservation reasons (O'Brien 2009: 14). The arrival of these vessels was labelled an 'invasion' (Anon 2008: 9, see also New Quay News 2008), as small-scale local fishermen contested the right of those from outside the area to fish there.

With regard to environmental concerns, the 2009 consultation aimed for additional regulation of the fishery, doing this in line with the *Welsh Fisheries Strategy*, in developing an 'ecosystem based approach' (WAG 2008: 2) on the basis of 'further research' (WAG 2008: 5). Responses to the consultation were, in many ways, unsurprising: the Marine Conservation Society saw it as an opportunity 'to make decisions based on sound scientific evidence and apply the precautionary principle wherever this knowledge is lacking' (WAG 2009: 2); a scallop fisherman commented that 'there was a startling lack of science involved in the formation of the Cardigan SAC' (WAG 2009b: 12); and another scallop fisherman felt that the

Cardigan Bay fishery 'is clearly sustainable' as it is 'the only fishery [in Wales] …
to withstand effort and is as good today as was 30 years ago' (ibid). Others felt
that, whatever the basis for new regulations it would 'be very difficult to assess …
what has and has not been successful' (WAG 2009b: 8).

The peak of the debate might be seen as the submission of complaints to the
European Commission at the end of the 2009 season over the possibility that WAG
was 'in breach of EU law by failing to undertake an Appropriate Assessment for
the Cardigan Bay SAC' (Save Our Sea 2009) in relation to the potential harm
caused to the listed habitats and species by dredging. As a result, WAG extended
the close season to 1 March 2010, to 'allow sufficient time to gather and process
additional scientific information in relation to the features of the SAC' (Hinz et al.
2010a: 2). The debate reached at least a temporary conclusion in 2010, with the
introduction of the Scallop Fishing (Wales) (No.2) Order 2010. This instituted a
new close season from 1 May to 31 October, along with various restrictions on
engine power, a prohibition of dredging within 1nm of the coast, and a prohibition
of scallop fishing in the 1–3nm zone unless the boat is less than 10 metres and
has a maximum of six dredges, in the 3–6nm zone unless it is towing no more
than 8 dredges, and in the 6–12nm zone unless towing no more than 14 dredges
(The Scallop Fishing (Wales) (No. 2) Order 2010: 5–6). Part of Cardigan Bay was
closed to dredging, although this continues in part of the SAC.

This section has outlined key debates around, and events in, the Cardigan
Bay scallop fishery. While state regulation forms the basis for management,
this was until recently implemented by a range of regional representatives from
political and environmental organizations. Although the fishery has been brought
more directly under state regulation once more, non-state organizations continue
to heavily influence the decisions made around it. However, this section has
portrayed the debate largely as focused on the politics of managing effort around a
declining stock, and of limiting the impact of particular fishing technologies. The
fish, technologies and other non-humans implicated in the debate are conspicuous
by their silence, a silence that I address through the remainder of the chapter.
I argue in the next section that both the debate around scallop fishing, and the
practice of management itself, are co-produced by the non-humans, and that their
actions disturb and contest the ways they have been represented.

Engaging the More-Than-Human: Dissidence

A key controversy in the debate around the scallop fishery regarded the extent to
which large dredgers could be held responsible for apparent long-term damage to
the seabed. Such alleged damage had received widespread publicity throughout the
region, particularly through the local press, which accused the dredgers of reducing
'seabed … to a featureless desert of mud and sand' (Anon 2008: 22), and through
the Save Our Sea campaign. The website of the latter featured images of dredge
'tracks' on the seabed, stating that 'many … organisms are killed, let alone disturbed

by dredging' (Save our Sea 2009) and that dredging 'leaves a barren landscape' (Save Our Sea 2009). This characterization of a stable and fixed ecosystem being targeted by mobile technologies was further emphasized by Monbiot (2009):

> The boats are not resident here. They move around the coastline trashing one habitat after another. They will fish until there is nothing left to destroy, then move to the next functioning ecosystem. If, in a few decades, the scallops here recover, they'll return to tear this place up again.

Such evidence was viewed as anecdotal, however, and for WAG was not suitably convincing to institute a ban on the practice. In one sense, then, the controversy around the fishery developed through scientific uncertainty. As a result, during the extended close season in 2009–2010, WAG commissioned additional research to directly assess the impact of dredgers. A first survey, relying on a combination of sonar and camera studies, was completed in December 2009, reporting in January 2010, finding 'no physical traces of scallop dredging on the seabed' (Hinz et al. 2010a: 24). The authors of the study also noted, however, that it was not possible to state definitively, in such a short study, that this meant scallop dredging does *not* cause damage, noting that the seabed is 'highly mobile' and thus any damage may have been hidden temporarily by shifting sand and suchlike (Hinz et al. 2010a: 25). As a result, a follow-up study was commissioned; this stated that 'natural processes outweighed the negative effects associated with scallop dredging within this highly dynamic area' (Hinz et al. 2010b: 29). The conservationists have again contested these findings, Friends of Cardigan Bay (Hughes 2011: 21) stating that they 'remain unconvinced of this and suggest that further research is required. It can take many years for an area to regenerate after the heavy and intrusive damage that scallop dredging inflicts.'

Here, the actions of the sea, currents and the material that composes the seabed might be understood to contribute directly to the debate around fishery management in the Bay. Through their images of dredging tracks, the conservation groups have illustrated a construction of dredging technologies as negatively impacting on a helpless and innocent natural environment. The scientists, however, have 'defined and distributed' (Woods 1998: 323) the roles of actors quite differently, showing through their study that the seabed's mobility matters. As such, following Callon (1986: 223), the scientists might be understood as successfully mobilizing a network, establishing themselves as 'spokesm[e]n' and, in relative terms, 'silencing' other 'problematizations' (Callon 1986: 203) of the situation.

Equally, the conservationists' approach might through this analytical frame be understood as less successful, failing to fully mobilize a network (though they undoubtedly had some success in promoting debate in the first place and in promoting a precautionary approach 'until further habitat surveys can confirm that scalloping is not adversely damaging' the SAC [Marine Conservation Society 2010]). While the images and commentary on their website presented pictures of a stable seabed damaged by the mobile technologies, their 'representivity [was] brought into question' (Callon 1986: 220). As such, the interrelationships of the sea, currents and seabed themselves disrupted the network that the conservationists had attempted to establish.

In one sense, this analysis focuses attention on the actions of non-humans and points to a more heterogeneous understanding of the actors involved in fisheries management. In another sense, the role of the non-humans here remains somewhat two-dimensional, with their actions merely disrupting human representations. The next section develops this theme further by highlighting the role played by non-human *movement*, not only in disrupting or emphasizing representations but in establishing the framework for regulation. Movement, in other words, may be a more constructive way through which to conceptualize these more-than-human practices of fisheries governance.

Engaging the More-Than-Human: Movement

In the previous section, non-humans were introduced as actants in the debate around impacts on the fish, dolphins and seabed. Through this, the materiality of the sea itself became a problematic in the dispute, raising questions not only about the extent to which dredgers might be seen as the prime cause of damage, but also about the authority of the initial scientific study, which viewed the seabed's constant change as an obstacle to the achievement of a definitive conclusion, rather than as a necessary attribute of an emergent set of relations. The second scientific report viewed this emergence more constructively, though this destabilized the conservationists' arguments in the process.

Emergence was also a key feature in a second area of debate, though here it involved the movement of actors in and out of the Bay, introducing elements of absence, presence and rhythmicity. This concerned the relationship between dolphin and scallop habitat. Dolphins entered the debate because of their dependence on a similar habitat to the scallops. According to the Countryside Council for Wales (CCW), the greatest proportion of dolphin diet in the Bay is composed of demersal species, such as cod, which 'live close to the seabed and are reliant upon it for food and shelter' (Taylor 2010: 8). In other words, while these species are not in a symbiotic relationship, damage to scallop habitat is likely to also affect the food supply of dolphins.

However, while adult scallops are relatively sedentary in the Bay, as previously noted, and are ever-present, the same cannot be said of the dolphins, which swim some distance. One response to the 2009 consultation commented on this, questioning whether a ban on scallop dredging in the SAC would help dolphins, given their absence 'during the winter months' (which largely coincides with the scallop fishing season [WAG 2009b: 10]). This movement considerably complicates the regulation of the fishery. As Lulka (2004: 439) has observed, the actions of environmental management often attempt to stabilize spatial structures in order to allow 'social and economic practices ... to continue relatively unimpeded' by the movement of animals. Indeed, in the case of Cardigan Bay, Feingold et al. (2010: 1) note that 'The boundaries of the current SAC were based on early studies. Since then, further seasonally important areas in the region have been identified'. This research suggested the need for further work to be carried out on the dolphin's migratory habits, but the implication is that the current SAC boundaries do not sufficiently account for the movement that takes place in and around them.

However, while the Bay is in many ways treated as a physically and geographically stable entity, the movements of dolphins, and their interactions with various fish species, force an expanded regulatory framework, which extends considerably beyond Cardigan Bay itself; because of the dolphin's movements, CCW also recommended that scallop dredging should not take place in part of Conwy Bay, in north Wales, as it 'is regularly used by bottlenose dolphins for feeding and the dolphins using the area are part of the Cardigan Bay population, and thus a feature of the Cardigan Bay SAC'.

Because of the movement of dolphins, therefore, there is tension between the stabilization and extension of spatial management boundaries. In some ways, the management framework remains relentlessly Euclidean, relying on surprisingly angular and rigid boundaries, but in others it must account for the mobile actions of the actants it attempts to protect or control. While in the previous section the non-human actants destabilized particular social constructions through their dynamism and change, here the contribution might be seen as more constructive. While the dolphins' movements, and the changing understandings of these, continue to destabilize attempts at management, the creatures are active in co-producing new spaces of management, linking geographically distant places.

Conclusions: More-Than-Human Fisheries Governance

This chapter has demonstrated two of the ways in which it would be inappropriate and misleading to characterize the debate around scallop dredging in Cardigan Bay as being merely about *human* economics and politics. Non-human actants, whether animals or non-sentient matter, are not merely passive objects, controlled and utilized by humans, but are active in co-producing new management frameworks and in aiding the contestation of particular representations. As such, the examples have begun to move some way towards Latour's vision of a heterogeneous collective, in which *a priori* distinctions are not drawn between 'nature' and 'society', the focus instead turning to activity, interaction and relations.

In *Politics of Nature*, Latour aims 'to produce new collectives of facts, values, and practices that will allow plural actors – humans and non-humans – to speak about common 'matters of concern' (Wainwright 2005: 116). As noted previously, key to Latour's thesis is the notion of 'articulation'. Here, articulation is not merely about vocalization, though that may be an important feature. In the heterogeneous collective, it is 'neither nature nor humans' who speak, 'but *well-articulated actors*, associations of humans and nonhumans' (Latour 2004: 86 original emphasis).

Fisheries regulation and management are increasingly characterized by a governance approach, which allows for the inclusion and participation of a greater range of actors than a conventional state-led government approach. The challenge raised by this chapter is for future fisheries governance to be more attentive to the voices of others than humans. As St Martin and Hall-Arber (2008: 779) have noted, the development of geospatial technologies increasingly make 'visible what had previously been hidden or inaccessible. Living and mineral resources, marine habitats, environmental conditions, sea bottom morphology, and species ranges and interactions are all increasingly documented and mapped.' While these authors comment on the relative sidelining of human interests in the application of such technologies, this chapter points towards the possibility of more explicitly involving non-humans in the governance process, listening to the dialogues that are produced and being attentive to their movements and interactions. The rigid configuration of the Special Area of Conservation in Cardigan Bay is at odds with the actants it aims to control and protect, and the brief examples provided in this chapter imply that a more nuanced system, which attends to seasonality and movement, would more symmetrically and inclusively involve a greater range of actants. Such a move is increasingly practically possible, as outlined by Martin and Hall-Aber (ibid). The challenge is to use these technologies constructively, allowing management practices to continue to adapt as the relationships change.

Finally, this chapter differs from the majority of scholarship on the ocean, acting as a counter to the frequently anthropocentric focus of much research, analysing it instead as a lively space, accounting for its fluid materialities and more-than-human composition. As such, it takes the sea and its multiple materialities seriously. Here, the sea itself is far from objectified, a passive space through and over which humans pass. It is dynamic and active, immanent and changing. Future research would usefully further engage with this dynamism, giving a fuller and deeper account of sea fisheries and their management.

References

Anon. 2008. Invasion fears. *Cambrian News*, 13 November, 9.

Anon. 2010. *Explanatory Memorandum to the Scallop Fishing (Wales) Order 2010*, Cardiff: National Assembly for Wales, Department for Rural Affairs.

Bear, C. 2011. Being Angelica? Exploring individual animal geographies. *Area*, 43(3), 297–304.

Bear, C. and Eden, S. 2011. Thinking like a fish? Engaging with nonhuman difference through recreational angling. *Environment and Planning D: Society & Space*, 29(2), 336–52.

Callon, M. 1986. Some elements of a sociology of translation – domestication of the scallops and the fishermen of St. Brieuc Bay. In *Power, Ation and Belief: A New Sociology of Knowledge?*, edited by Law, J. London: Routledge and Kegan Paul, 196–233.

Cardigan Bay SAC. Date unknown. *Bottlenose Dolphin*. [Online] Available at: http://www.cardiganbaysac.org.uk/?page_id=72 [accessed: 4 June 2012].

Castree, N. and Braun, B. (eds), 2001. *Social Nature: Theory, Practice, and Politics*. Oxford: Blackwell.

Ceredigion County Council, Countryside Council for Wales, Environment Agency Wales, North Western and North Wales Sea Fisheries Committee, Pembrokeshire Coast National Park Authority, Pembrokeshire County Council, South Wales Sea Fisheries Committee, and Dwr Cymru Welsh Water. 2008. *Cardigan Bay Special Area of Conservation (SAC) Management Scheme*. Aberaeron: Ceredigion County Council.

Constance, D.H. and Bonanno, A. 2000. Regulating the global fisheries: The World Wildlife Fund, Unilever, and the Marine Stewardship Council. *Agriculture and Human Values*, 17(2), 125–39.

De Vivero, J.L.S., Mateos, J.C.R. and del Corral, D.F. 2008. The paradox of public participation in fisheries governance: The rising number of actors and the devolution process. *Marine Policy*, 32, 319–25.

Eden, S. and Bear, C. 2010. Third-sector global environmental governance, space and science: Comparing fishery and forestry certification. *Journal of Environmental Policy & Planning*, 12(1), 83–106.

Evans, J. 2011. *Environmental Governance*. London: Routledge.

Feingold, D., Baines, M. and Evans, P.G.H. 2010. *Cardigan Bay Bottlenose Dolphin Social and Population Structure – Findings from a Ten-Year Photo ID Dataset* [Online]. Available at: http://www.seawatchfoundation.org.uk/docs/Daphna%20Feingold_poster_11_final.pdf [accessed: 15 June 2012].

Gibbs, M.T. 2008. Network governance in fisheries. *Marine Policy*, 32, 113–19.

Gullo, A., Lassiter, U. and Wolch, J. 1998. The cougar's tale. In *Animal Geographies: Place, Politics and Identity in the Nature-Culture Borderlands*, edited by J. Wolch and J. Emel. London: Verso, 139–61.

Hatanaka, M., Bain, C. and Busch, L. 2005. Third-party certification in the global agrifood system. *Food Policy*, 30(3), 354–69.

Hinchliffe, S. 2008. Reconstituting nature conservation: Towards a careful political ecology. *Geoforum*, 39(1), 88–97.

Hinchliffe, S. 2010. Working with multiples – a non-representational approach to environmental issues. In *Taking Place: Non-representational theories and geography*, edited by B. Anderson and P, Harrison. Farnham: Ashgate, 303–20.

Hinchliffe, S., Kearnes, M.B., Degen, M. and Whatmore, S. 2005. Urban wild things: A cosmopolitical experiment. *Environment and Planning D: Society & Space*, 23(5), 643–58.

Hinz, H., Scriberras, M., Benell, J.D. and Kaiser, M.J. 2010. *Assessment of Offshore Habitats in the Cardigan Bay SAC*. Bangor: Bangor University.

Holloway, L. 2007. Subjecting cows to robots: Farming technologies and the making of animal subjects. *Environment and Planning D: Society & Space*, 25(6), 1041–60.

Hughes, P. 2011. Scallop dreding in Cardigan Bay – an update. *Friends of Cardigan Bay Newsletter*, 2010/11, 18–21.

Johnston, C. 2008. Beyond the clearing: towards a dwelt animal geography. *Progress in Human Geography*, 32(5), 633–49.

Jones, O. 2000. (Un)ethical geographies of human-animal realtions: Encounters, collectives and spaces. In *Animal Spaces, Beastly Places: New Geographies of Human-Animal Relations*, edited by C. Philo and C. Wilbert. London: Routledge, 268–91.

Latour, B. 1993. *We Have Never Been Modern*. Cambridge: Harvard University Press.

Latour, B. 2004. *Politics of Nature*. Cambridge: Harvard University Press.

Latour, B. 2005. *Reassembling the Social*. Oxford: Oxford University Press.

Lorimer, H. 2005. Cultural geography: The busyness of being 'more-than-representational'. *Progress in Human Geography*, 29(1), 83–94.

Lorimer, H. 2006. Herding memories of humans and animals. *Environment and Planning D: Society & Space*, 24(4), 497–518.

Lorimer, J. 2008. Counting corncrakes: The affective science of the UK corncrake census. *Social Studies of Science*, 38(3), 377–405.

Lorimer, J. 2010. Moving image methodologies for more-than-human geographies. *Cultural Geographies*, 17(2), 237–58.

Lulka, D. 2004. Stabilizing the herd: Fixing the identity of nonhumans. *Environment and Planning D: Society & Space*, 22(3), 439–63.

Lulka, D. 2009. The residual humanism of hybridity: Retaining a sense of the earth. *Transactions of the Institute of British Geographers*, 34(3), 378–93.

Mansfield, B. 2004a. Neoliberalism in the oceans: "Rationalization," property rights, and the commons question. *Geoforum*, 35(3), 313–26.

Mansfield, B. 2004b. Rules of privatization: Contradictions in neoliberal regulation of North Pacific fisheries. *Annals of the Association of American Geographers*, 94(3), 565–84.

Mansfield, B. 2007a. Privatization, property and the remaking of nature-society relations: Introduction to the special issue. *Antipode*, 39(3), 393–405.

Mansfield, B. 2007b. Articulation between neoliberal and state-oriented environmental regulation: Fisheries privatization and endangered species protection. *Environment and Planning A*, 39(8), 1926–42.

Marine Conservation Society. 2011. *Wales Shows The Way With The New Welsh Scallop Order – Time For The Rest Of The UK To Take Note* [Online]. Available at: http://www.mcsuk.org/press/view/288 [accessed: 8 February 2010].

Matless, D., Merchant, P. and Watkins, C. 2005. Animal landscapes: Otters and wildfowl in England 1945–1970. *Transactions of the Institute of British Geographers*, 30(2), 191–205.

Meade, G. 2012. Reforms 'will not stop overfishing'. *Independent* [Online]. Available at: http://www.independent.co.uk/environment/nature/reforms-will-not-stop-overfishing-7844086.html [accessed: 12 June 2012].

Monbiot, G. 2009. These are not the mariners of old but pirates who make bureaucrats blanch. *Guardian* [Online]. Available at: http://www.guardian. co.uk/commentisfree/2009/jun/01/george-monbiot-marine-fisheries-law [accessed: 10 May 2010].

New Quay News [Online]. Available at: www.new-quay.com/n6nov8.htm [accessed: 4 July 2010].

North Western Inshore Fisheries and Conservation Authority. Date unknown. *NW&NWSFC Archive – Committee* [Online]. Available at: http://www. cumbriasfc.org.uk/ContentDetails.aspx [accessed: 20 June 2012].

O'Brien, P. 2009. Scallop boats with low-paid crews 'threaten local jobs'. *Cambrian News*, 26 February, 14.

Peters, K. 2010. Future promises for contemporary social and cultural geographies of the sea. *Geography Compass*, 4(9), 1260–72.

Philo, C. 1995. Animals, geography, and the city: Notes on inclusions and exclusions. *Environment and Planning D: Society & Space*, 13(6), 655–81.

Philo, C. 1998. Animals, geography and the city: Notes on inclusions and exclusions. In *Animal Geographies: Place, Politics, and Identity in the Nature-Culture Borderlands*, edited by J. Wolch and J. Emel. London: Verso. 51–71.

Philo, C. 2005. Spacing lives and lively spaces: Partial remarks on Sarah Whatmore's 'Hybrid geographies'. *Antipode*, 37(4), 824–33.

Philo, C. and Wilbert, C. 2000. Animal spaces, beastly places: An introduction. In *Animal Spaces, Beastly Places: New Geographies of Human-Animal Relations*, edited by C. Philo and C. Wilbert. London: Routledge, 1–34.

Save Our Sea. 2009. *Save Our Sea* [Online]. Available at: www.savecardiganbay. org.uk [accessed: 23 June 2012]. Sea Fisheries – *The North Western and North Wales Sea Fisheries District (Consequential and Transitional Provisions) Order 2010*. SI 2010/631. London: HMSO.

St Martin, K. 2006. The impact of "community" on fisheries management in the U.S. Northeast. *Geoforum*, 37(2), 169–84.

St Martin, K. and Hall-Arber, M. 2008. The missing layer: Geo-technologies, communities, and implications for marine spatial planning. *Marine Policy*, 32(5), 779–86.

Steinberg, P.E. 2001. *The Social Construction of the Ocean*. Cambridge: Cambridge University Press.

Taylor, J. 2010. *Scallop Dredging – Letter of Advice from Countryside Council for Wales to the Welsh Assembly Government* [Online]. Available at: http://www. ccw.gov.uk/about-ccw/newsroom/press-releases/idoc.ashx?docid=1626b5ec-bbed-4807-b008-6073e0f22a54&version=-1 [accessed: 6 October 2011].

Wainwright, J. 2005. Three recent works by Bruno Latour. *Capitalism Nature Socialism*, 16(1), 115–27.

Welsh Assembly Government. 2008. *Wales Fisheries Strategy*. Cardiff: Welsh Assembly Government.

Welsh Assembly Government. 2009a. *Proposals for the Scallop Fishery in Wales*. Cardiff: Welsh Assembly Government.

Welsh Assembly Government. 2009b. *The Scallop Fishery in Wales – Consultation 17ᵗʰ July–25ᵗʰ September 2009: Summary of responses*. Cardiff: Welsh Assembly Government.

Whale and Dolphin Conservation Society. 2011. *Press release: Welsh Assembly Restriction on Scallop Dredging doesn't go Far enough Says Whale and Dolphin Conservation Society* [Online]. Available at: http://www.cisionwire. com/whale-and-dolphin-conservation-society/r/welsh-assembly-restriction-on-scallop-dredging-doesn-t-go-far-enough-says-whale-and-dolphin-conservation-society,c470881 [accessed: 5 February 2010].

Whatmore, S. 2006. Materialist returns: Practising cultural geography in and for a more-than-human world. *Cultural Geographies*, 13(4), 600–609.

Wolch, J. 2002. Anima urbis. *Progress in Human Geography*, 26(6), 721–42.

Wolch, J. and Emel, J. 1995. Bringing the animals back in. *Environment and Planning D: Society & Space*, 13(6), 632–6.

Woods, M. 1998. Researching rural conflicts: Hunting, local politics and actor-networks. *Journal of Rural Studies*, 14(3), 321–40.

Chapter 11

'With perfect regularity throughout': More-Than-Human Geographies of the Royal Mail Steam Packet Company

Anyaa Anim-Addo

Introduction

Steamship services connected imperial, colonial and extra-imperial sites during the nineteenth century, facilitating 'the movement of capital, people and texts' around the world (Lambert and Lester 2006: 10). Steamships can thus be understood as one of the nineteenth century's key 'technologies that enhance the mobility of some peoples and places even as they also heighten the immobility of others' (Hannam, Sheller and Urry 2006: 3). A company that represents well such technology and corresponding mobility is the Royal Mail Steam Packet Company (RMSPC), which connected Britain with the Caribbean and South America. The RMSPC sought to privilege the circulation of imperial news and correspondents, and the movements of 'imperial careerists' (Lambert and Lester 2006: 2). At the same time, the RMSPC enabled various unintended trans- and circum-Atlantic mobilities. In this chapter (following Bear, this collection), I use the case study of the RMSPC to examine the significance of more-than-human geographies in the maritime world.

To contextualize, the RMSPC was an imperial project envisaged by James MacQueen, who planned an ambitious steamship service that would link eastern and western parts of the British Empire (MacQueen 1838). MacQueen, who had worked as a manager of a sugar estate during the late eighteenth century, went on to become a vocal critic of abolitionists during the 1820s and 1830s (Lambert 2008). Having known and made a living out of the British Caribbean under slavery, MacQueen hoped that steamship communication between Britain and the region would mitigate post-emancipation instabilities, particularly by promoting commerce. In a letter addressed to Francis Baring, MP, MacQueen argued in favour of his scheme, and wrote that, '[s]tartling as the subject of connecting China and New South Wales with Great Britain, through the West Indies, may at first sight appear, both as regards time and expense, still few things are more practicable' (MacQueen 1838: 5–6). Prior to the RMSPC's service, mail was transported across the Atlantic under sail from the packet station at Falmouth. Larger West Indian islands received their mail from brigs converted to steam (Bushell 1939: 3–4).

MacQueen explained that within his envisaged scheme, the West India station would be 'one of the most important, and extensive, and complicated of the whole, and one where steam-vessels [could] be employed with the most beneficial effects' (MacQueen 1838: 28). A scaled-down version of MacQueen's plan, initially only serving the West Indies, was funded by the British Government in the form of a mail contract subsidy, worth £240,000 a year. Similarly to P&O, another mail-contract holding line, the RMSPC sought to present itself 'as existing not merely for the good of its shareholders but for the good of the nation' (Harcourt 2006: 3).

Steamship lines such as the RMSPC secured public funding precisely because they promised predictable and timetabled, or 'regular' service, and thus regular communication with colonies and extra-imperial spaces. The concept of regularity was invested with great significance by RMSPC managers and directors. Yet, it should be noted that despite the introduction of steamship technology, the RMSPC's vessels in the 1840s were hybrid in so far as they relied on steam power but could also harness sail power when necessary. When the Company's complicated scheme of routes proved over-ambitious during the first few months of operations in 1842, the directors reported their 'regret' that 'the service at its commencement was not performed with the regularity that could have been wished' (UCL RMSP 13, 28 September 1842: 1). Two years later, by contrast, the Company wrote to the British Admiralty on the subject of its modified scheme of routes, which had been in operation for six months. The RMSPC reported with 'great satisfaction' that the ships had 'performed their voyages with perfect regularity throughout' (NMM RMS 7/1, 8 January 1844). The report of 1846 stated that '[t]he Company's Ships continue to arrive and depart at the various places, both abroad and at home, with the greatest regularity, and the manner in which the Service is conducted appears to afford general satisfaction' (UCL RMSP 13, 15 October 1846: 1). Crosbie Smith, Ian Higginson and Philip Wolstenholme's work has highlighted the strategies deployed by Alfred and Philip Holt in building credibility for the Liverpool-based Ocean Steamship Company. The Holts' relationships of trust within engineering networks were framed by 'a moral economy, minimizing true waste and maximizing useful, *qua* good, work' (Smith, Higginson and Wolstenholme 2003: 388). The RMSPC sought to build credibility partly by investing in a rhetoric of regularity when communicating with the British Government and the public. This concept of 'regularity' was tied to notions of order and efficiency, and the Company sought to order its ships socially and materially, desiring them to be places of 'order, cleanliness, and efficiency' (NMM RMS 38/1: 172). The Company's vessels were highly regulated spaces, with hierarchies, divisions and surveillance mechanisms working to order journeys, Company infrastructure and the timetable. As Frances Steel notes, steamship services 'fostered a stronger, more confident articulation of industrial achievement with national strength and imperial influence', and the notion of regularity was key to this expression (Steel 2011: 4). As I will suggest in the next two sections of this chapter, idealized notions of order were frequently challenged, and in this respect, more-than-human factors had a significant impact on maritime operations.

In his article, 'Towards a politics of mobility', Tim Cresswell (2010) outlines six constituent parts of mobility. He suggests that the first of these is motive force, which causes a person or thing to move. The second is velocity, rhythm is a third important factor in mobility, and route is a fourth element in the equation. The experience of mobility, or what it feels like, is a fifth aspect. Lastly friction, which slows down or causes movement to stop, is to be considered the final facet of mobility. Cresswell indicates that these different elements combine to create 'constellations of mobility' at particular times (Cresswell 2010: 17). In this chapter I draw on Cresswell's disaggregation of mobility to explore how more-than-human geographies affected the velocity, rhythms, routes and experience of the RMSPC's shipping service, and even, on extreme occasions of 'friction', disrupted the service altogether. I examine the ordinary and exceptional ways in which the (ir)regularity of the steamship service was co-produced by more-than-human geographies, in the light of which I argue that during this period, a strong dissonance existed between the rhetoric of regularity attached to steamship transport, and the significance of more-than-human geographies in shaping steamship journeys.

Hybrid, posthuman and more-than-human theoretical approaches have questioned definitions of the human subject, and have highlighted the 'complexity and interconnection of life' (Panelli 2010: 80, see also Castree and Nash 2006, Coyle 2006, Whatmore 2002). Sensitivity to more-than-human actants is particularly appropriate to maritime research since, as Leah Gibbs (2009: 361) suggests, 'water places' reveal the 'complex interactions that comprise a more-than-human world'. This chapter approaches such interactions through the historical case study of the RMSPC's operations in the nineteenth-century Caribbean. As I will suggest, the Company's service comprised of a network of 'diverse objects, organisms, forces and materialities' (Lorimer 2010: 238). Thus the RMSPCs commitment to providing a regular and timetabled service strained against the messy '"hydro" materiality of the sea' (Peters 2012: 1242).

In the field of Caribbean studies, Bonham Richardson has critiqued the historiographical absence of work that pays attention to 'environmental considerations' (Richardson 1997: 13). The focus of this chapter upon the Caribbean's historical maritime places addresses this concern to bring to the fore the significance of environmental and other more-than-human factors in the region's past. Indeed, an examination of more-than-human maritime historical geographies provides one important way to further challenge 'the longstanding separation of urban and environmental studies', since port towns as well as shipping services depend on more-than-human interactions in coastal and deep-sea spaces (Braun 2005: 635). Further, while the advent of steamship technology was lauded in the nineteenth century as a means to end reliance on sail power and to sever journeys from natural rhythms, the RMSPC's operations between the 1840s and the 1860s reveal that early steamship technology continued to be shaped by more-than-human 'entanglements of people, animals and technologies' (Castree and Nash 2006: 501–2).

In this chapter I argue that despite the Company's heavy investment in a discourse of regularity, efficiency and order, the RMSPC's service can be understood as a network 'of actants-in-relation that are at once local and global, natural and cultural, and always more than human' (Whatmore 1999: 33). Having highlighted the importance of notions of regularity and efficiency to nineteenth-century steamship services, in the next section, 'Mundane more-than-human geographies of the RMSPC', I examine the everyday ways in which the non-human world shaped steamship operations. In the final section of the chapter, 'Exceptional more-than-human impact', I explore extraordinary moments when the more-than-human world played a particularly prominent role in shaping the Company's service.

Mundane More-than-Human Geographies of the RMSPC

It is notable that as well as Company aspirations for a predictable and regular service, the British Admiralty also expected ships to move in strict adherence to the scheme of routes. Thus in response to the Company's superintendent at Grenada altering the timing of route number two (Grenada – Trinidad – Grenada – Barbados – Grenada) in the Lesser Antilles, the Admiralty indicated in unequivocal terms that the scheme 'should not be deviated from or experimented upon' (NMM RMS 6/3, 13 January 1844). Yet despite Admiralty expectations, the scheme of routes proved to be dynamic. Thus between January and December 1844, the RMSPC deviated from the printed scheme on nine separate occasions. Ships were delayed, for example at Havana in April 1844, but vessels also left their destinations too early (without waiting for the connecting line), as occurred in the same month at Nassau. Thus the Admiralty's desire for calculable and perfectly regular travel failed to correspond with the working realities of an intricately interconnecting service (NMM RMS 7/2, 4 March 1845). Even those familiar with the West Indies were inexperienced in piloting large steamships through those waters. Imperfect knowledge of the Caribbean seascape resulted in the grounding of ships, and therefore delays. Thus the scheme of routes was under frequent negotiation between humans. In addition to this, the timetable also operated through an important kind of negotiation with the more-than-human world. In this way, the steamship service was significantly shaped by 'the temporal rhythms of human/non-human difference' (Whatmore 2002: 4–5).

Despite ordered ideals, human and more-than-human factors repeatedly undermined the regularity of the RMSPC's service. Firstly, human errors proved routine. For example, the RMS *Thames* became grounded and was delayed in April 1844 due to the unexpected absence of a local pilot (NMM RMS 7/2, 15 July 1844). On another occasion, the commander of the schooner *Liffey* fell ill and the chief mate of the *Tay* was redeployed to assist the *Liffey*, leaving the *Tay* short of officers (NMM RMS 7/2, 3 October 1844). Moreover, despite the existence of extensive rules designed to order the ship, experience brought to light officers' ignorance of Company regulations. Thus the RMSPC deemed it necessary to request from 1850 onwards that commanders examine officers on their knowledge of these (NMM RMS 38/1: 19). The Company's ordering strategies worked against the counter-current of human idiosyncrasies and irregularities. Simultaneously on board any given ship, the directors' and managers' desire for order strained against individuals' tendencies to act irregularly, spontaneously, or in contradiction to the Company's regulations.

Essentially, the RMSPC's desire to run an ordered and regular service was belied by the tendency of the world to 'kick back' against the neat spatio-temporal mapping of the Company's scheme of routes (Whatmore 2002: 4–5, Barad 1998). As Steel (2011: 5) has highlighted, steamship operations 'were always messy'. I suggest that part of this messiness derived from the more-than-human geographies that shaped these logistical operations. The more-than-human world affected the RMSPC's service in various everyday ways, and particularly influenced the timetable. On one occasion, when the RMS *Thames* was travelling through the Caribbean in 1844, extreme bad weather at Bermuda meant that the ship arrived at Havana two days behind schedule (NMM RMS 7/1, 15 July 1844). During another journey, the RMS *Tay* arrived off the Bar of Tampico on 17 January 1845, but a violent northerly wind forced the ship to take shelter. It was only when the weather calmed on 20 January that the *Tay* was able to land the mail at Tampico (NMM RMS 7/1, 3 April 1845). Similarly, when the RMS *Avon* left England with the mail in December 1848, the ship experienced such 'tempestuous weather' in the early part of its journey that the captain and Admiralty Agent decided to alter the normal course and to coal at Madeira before steaming straight to St Thomas (NMM RMS 7/2, 15 February 1849).[1] As a result, the delivery of the mail was severely delayed at multiple destinations in the Caribbean. Meteorological factors shaped RMSPC navigation, which in turn affected the rhythms of the service, presenting a challenge to the regularity that the Company sought to ensure. In this respect, the RMSPC's officers and crew were faced with frequent reminders that they were 'powerless to achieve their desired outcomes unless the non-human entities perform[ed] the roles ascribed to them' (Woods 2007: 498).

The RMSPC's attempts to secure the integrity of the mail on board equally indicates the significance of more-than-human considerations. Firstly, post had to be kept dry, and the mail room needed to be secured against leaks, as, for example, a leaky mail room threatened to destroy the post on the RMS *Trent* in 1844 (NMM RMS 6/3, 1844–1845). During the first decade of service, the Company also found that rats had a tendency to destroy the mail bags and damage the post. In response, the Company hired a rat-catcher to reduce the effects of this animal geography when the vessels were in Southampton waiting to depart, and further ruled in 1851 that cats would be sent to sea on the ships (NMM RMS 38/1: 7). Whereas the RMSPC sought to regulate the rats' incursion into the space of the ship, cats

---

1 The Admiralty Agent was the British Admiralty's representative on board the RMSPC's steamers, and was responsible for overseeing the safe delivery of the mail.

became part of the regulatory infrastructure of the ship space. In this way, the RMSPC's service was shaped by a 'hybrid engagement of human and non-human entities at all scales' (Woods 2007: 487).

While animal geographies affected the integrity of the post, a starkly contrasting perspective on human/ animal interactions on board ship is provided through Charles Kingsley's account of a journey to the West Indies on board an RMSPC steamer. His writing indicates that passenger impulses to collect as they travelled meant that vessels were also at times deliberately internally constituted as more-than-human worlds for entertainment rather than functional reasons. Kingsley recalled that travelling on board the RMS *Neva*, under the command of Captain Woolward, the passengers and crew staged a 'wild-beast show' for their own entertainment. The doctor 'contributed an alligator' and the chief engineer exhibited a 'live Tarantula' (Kingsley 1874: 396). While Kingsley's claim of an alligator on board is startling, his narrative certainly suggests that animals and insects accompanying passengers sometimes served as part of the spectacle and entertainment of steamship travel, breaking up the monotony of oceanic journeys. Animal presence on board RMSPC steamers was thus shaped both by necessity and curiosity. The travelling steamers' interior spaces were constituted, intentionally and unwittingly, as more-than-human worlds.

Weather, water and animals were all co-constituents of the RMSPC's service – its integrity, its safety, its rhythms, and spaces. The RMSPC saw these as elements of its service to be ordered and regulated, while passengers occasionally sought to construct a more-than-human presence as a spectacle to be staged within the ship space. In these ways, more-than-human movements and interactions shaped the Company's operations on a weekly basis during the normal course of service. Yet, as I will explore next, on rare occasions, the more-than-human world dramatically and dynamically altered the RMSPC's operations in the Caribbean.

To highlight, I argue that despite the Company's heavy investment in a discourse of regularity, efficiency and order, the RMSPC's service can be understood as a network 'of actants-in-relation that are at once local and global, natural and cultural, and always more than human' (Whatmore 1999: 33). In the next section I explore extraordinary moments of service when the more-than-human world played a particularly prominent role in shaping the Company's service.

Exceptional More-than-Human Impact

The winter of 1866–1867 proved traumatic for the RMSPC on the occasion of a yellow fever outbreak, which blighted the steamship service with death and disease. The RMSPC's crew and shore-based employees suffered devastatingly high mortality at St Thomas in the Danish West Indies, the Caribbean hub of the transportation network. To understand the import of this health crisis for the RMSPC, it needs to be appreciated that twice a month, a steamer on the 'Atlantic route' departed from Southampton, carrying mail, passengers and low-bulk high-value cargo, and made a journey of 14 days and nine hours to St Thomas. At St Thomas, this steamer connected with various RMSPC branch routes departing from St Thomas and travelling towards Jamaica and Colon, Havana and Puerto Rico, the Windward and Leeward Islands, Santa Martha, Cartagena and Grey Town. After a fortnight's pause at St Thomas, and having replenished its supply of coal for the homeward leg, the transatlantic steamer would set off towards Southampton with people and mail bound for British shores (NMM RMS 36/4). However in January 1867, the Company learned that one of its vessels, the RMS *La Plata*, had sailed from St Thomas harbour on 31 December 1866 with more than 50 of the crew sick, and that 23 of these were now deceased (NMM RMS 6/17, 7 January 1867 and 16 January 1867). Also in January, George Gibon, staff commander and naval agent, wrote to the Company to report 'a most poisonous malaria in the atmosphere' of St Thomas. 'Most of the native labourers and the crews of the collieries employed there in discharging coals, and coaling the Company's mail packets', he wrote, had died (NMM RMS 6/17, 16 January 1867).

The yellow fever virus, considered here as more-than-human, is spread by the Aëdes Aegypti mosquito when the insect ingests diseased blood and subsequently injects the virus into a healthy person. The mosquito is mostly found in urban areas, and prefers to breed in standing water (Hays 2005: 179). Doctors typically recognized patients from the jaundiced appearance of their eyes and skin, black vomit, and a high fever that could cause delirium (Hays 2005: 180). During the nineteenth century, medical opinion was divided on the cause of the illness, with some attributing yellow fever to atmospheric, miasmatic and meteorological influences, and others stressing sanitation and hygiene (De Paolo 2006: 122). Despite the different emphases in medical thought, as Philip Curtin indicates, the standard response to yellow fever in the West Indies had been a mobile one, namely flight from the outbreak (1998: 64–5). Yet flight was not a viable solution for the RMSPC. As a British Government mail-contract holder, the Company had a commitment to maintain communication with the West Indies, and thus required its ships to continue to pass through either St Thomas or an equivalent site where the mail transfer could be made.

Once the impact of yellow fever on crew mortality became apparent, the RMSPC took an initial step towards managing the crisis by instructing its transatlantic steamers to transfer passengers, mail and cargo to the Company's inter-colonial branch vessels just outside of St Thomas harbour (NMM RMS 54/2, 24 April 1867). The RMSPC's managers also instructed vessels to coal at Jamaica rather than at St Thomas. Yet the deaths raged on. In light of the continuing mortality, a letter to the *Times* highlighted the Company's decision, over the years, to retain St Thomas as the main Caribbean transfer station (*Times*, 21 January 1866). The correspondent accused the Company of being 'morally guilty of murder'. This debate about the location of the Company's hub was politically charged, since some interest groups and individuals had long been invested in seeking to persuade the RMSPC to use a British colony for this purpose, rather than St Thomas. At the end of January 1867, the RMSPC took a second step towards managing the crisis and altered the routes of its service. The Company changed its transfer point (for the exchange of post, cargo and passengers) to Peter Island, approximately 25 miles from St Thomas, with coaling to take place at other stations along the vessels' routes such as Grenada (NMM RMS 6/17, 31 January 1867).

As the RMSPC's yellow fever outbreak indicates, the steamship network facilitated intended mobilities, such as the circulation of people, Government despatches and news, but the service also enabled unwanted mobilities, some of which, as in the case of yellow fever, stemmed from more-than-human movements and concentrations. The Aëdes Aegypti mosquito and the yellow fever virus shaped the RMSPC's operations in 1867, temporarily altering the routes of the service and slowing the ordinary rhythms of steamers as they paused in quarantine. When yellow fever or any other 'highly infectious distemper' was prevalent in America or the West Indies, the Privy Council could require every vessel that had called at an infected port to anchor at a specified place, where the vessel would be visited to ascertain the health of the crew before the ship was allowed to proceed to its port of destination (Booker 2007: 256). Thus quarantine regulations could force ships to pause for several days before being cleared to disembark passengers and cargo, and during the yellow fever outbreak, the RMSPC's ships were detained in quarantine at the Mother Bank on their arrival in Britain (NMM RMS 4/4, 25 October 1867: 6–7). As Tim Edensor points out, '[j]ourneys have a particular rhythmic shape' (Edensor 2010: 6). Following this line of thought, the outbreak of yellow fever altered the 'rhythmic shape' of steamship passages. The presence of yellow fever around the town and harbour of St Thomas not only brought death and illness to RMSPC employees, but also altered the transport network's routes and subjected the Company to critical public scrutiny and additional expense. Unsurprisingly, passengers were deterred from travelling through St Thomas on board the Company's steamers at this time (NMM RMS 4/4, 28 October 1868: 4). The impact of the more-than-human world on the Company's operations was interpreted by the RMSPC's directors as a 'visitation of providence' (NMM RMS 4/4, 25 October 1867: 3–5). Even so, while the crisis was articulated in religious terms, the Company's logistical response to yellow fever suggests that the RMSPC understood the health crisis as related to particular sites. Essentially, the Company responded to the outbreak by altering the routes and hub of its operations. Thus the Company changed the geographies of its steamship network in response to this particular more-than-human concern. As the yellow fever outbreak of 1866–1867 highlights, the RMSPC's operations both constituted and were constituted by more-than-human mobilities.

Just as the yellow fever crisis appeared to subside towards the end of 1867, another more-than-human geography powerfully affected the Company's operations. The RMS *Rhone* departed from Southampton on 2 October 1867 under the command of Captain T. Woolley with mail for the West Indies, and arrived at St Thomas on 14 October. Due to yellow fever concerns at St Thomas, the *Rhone* proceeded to Peter Island on 24 November. The vessel was due to begin its return journey to Britain with passengers and post on the 29th of the

Fig. 11.1 'Wreck of the Royal Mail "Rhone"', by William Frederick. Mitchell © National Maritime Museum, Greenwich, UK.

month ('The hurricane in the West Indies', *Times*, 8 November 1867 and 13 November 1867). On the morning of 29 October, the weather was 'threatening' and the wind was blowing from northward (*Times*, 22 November 1867). Just after midday, the wind calmed, and the *Rhone* began steaming towards the sea. However a hurricane gathered force, driving the *Rhone* onto a reef at Salt Island and wrecking the vessel. Passengers, who were lashed onto the deck, drowned (*Times*, 23 November 1867). Officers and members of the crew also lost their lives in the tragedy. The *Rhone* had 145 people on board, of whom fewer than 30 survived ('The hurricane in the West Indies', *Times*, 21 November 1867). The hurricane swept through the island of St Thomas, tearing roofs away from houses, destroying lives, shipping and infrastructure. Neighbouring Tortola was also struck. The RMS *Conway* was driven ashore at Tortola and dismasted, but fared better than the *Rhone* in so far as the *Conway* was later recovered. As the hurricane began, the RMS *Wye* attempted to get up steam and proceed to sea, but was wrecked on Buck Island. The *Times* reported of the *Wye*'s survivors that five of the 13 amongst the crew were 'whites, including the captain, the chief officer, and the boiler maker', providing a rare glimpse into the social composition of the RMSPC's inter-colonial crew (*Times*, 22 November 1867). The RMS *Derwent* was also thrown ashore at St Thomas during the hurricane (NMM RMS 4/4, 29 April 1868: 3–4).

The RMSPC was only one of several shipping companies with crew and vessels near to St Thomas and Peter Island that day, and other 'vessels in the harbour were either sunk, smashed to pieces, or driven ashore dismasted' (*Times*, 22 November 1867). The RMS *Douro* was 250 miles away from St Thomas during the hurricane, and an officer described the scene when the *Douro* arrived at St Thomas harbour on 30 October:

> First impressions underwent a sad change when we got sufficiently near to see the harbour strewn with wrecks, the lighthouse gone, and many houses roofless. A confused mass, near the middle of the harbour, built up of crushed hulls, broken spars, and loose cordage, was formed by the ship *British Empire*, lately out of England with 3,800 tons of coal for the use of the steamers of the Royal Mail Company. (*Times* 22 November 1867)

By the afternoon of 30 October, 292 bodies were reported to have been washed ashore and buried (*Times* 22 November 1867). The crisis deepened further when an earthquake struck St Thomas on 18 November. In addition to the human consequences of these non-human mobilities, the hurricane and earthquake had serious financial implications for the RMSPC, particularly as the Company insured its own vessels, and the share price fell as a result of the hurricane (*Times*, 8 November 1867; RMS 4/4, 29 April 1868: 20–21). The RMSPC was forced to replace and repair vessels, and also spent almost 500 pounds 'in clearing away the Ruins in the Harbour of St Thomas and repairing Walls' (RMS 4/4, 29 April 1868: 12–13). On 18 November, the *Times* reported the scene at Southampton in dramatic terms:

> The docks and the Royal Mail Company's offices have for some days past been hourly besieged by anxious inquirers, especially women, too many of whom are in the most lamentable state of uncertainty as to whether they are at this moment wives or widows; and the dejected countenances of most of those who are to be seen pacing rapidly to and from the docks, despite the bitterly cold cast wind, proclaim them to be bound on the same melancholy errand. (*Times* 18 November 1867)

At Southampton, a public subscription fund was set up for widows, orphans and those who suffered loss as a result of the hurricane (RMS 4/4, 29 April 1868: 12–13). The effects of the hurricane elicited responses at multiple sites of the RMSPC's network in the Atlantic world.

I have written elsewhere of how labour in post-emancipation Caribbean port towns, as in inland post-emancipation spaces, was shaped by contests over mobility, and of the ways in which shore-based maritime labourers were accused of demanding unreasonable wages and avoiding work (Anim-Addo 2011). In the aftermath of the hurricane, this discourse was seemingly heightened as James Lamb, H.M. Consul at St Thomas, condemned the black inhabitants of St Thomas, writing that they had 'behaved badly in the past few days', requesting what he termed 'exorbitant wages', and had failed to cooperate with 'the authorities'. Another report emanating from St Thomas on 5 November claimed that 1,600 dollars were offered to a gang of 100 men for a day of work 'but not one could be obtained, as the blacks have struck work' (*Times* 23 November 1867). S.D. Smith writes of the 1831 hurricane in St Vincent and notes that 'the enslaved population did not take advantage of the disaster to mount a challenge to white authority', however the reports received in the aftermath of the St Thomas hurricane suggest that contests over labour could be exacerbated in exceptional moments of more-than-human impact (Smith 2012: 97). Irrespective of the truth of Lamb's claims of non-cooperation, the circulation of such reports indicates that these exceptional moments of more-than-human force brought to the fore pre-existing social divisions and conflicts in the post-emancipation Caribbean.

The extreme more-than-human impact of the hurricane and earthquake of 1867 on the RMSPC's service in the Caribbean and in Britain rendered the Company's service distinctly irregular for a period of time, as lives were lost, damaged ships and infrastructure had to be repaired and the schedule was thrown into disarray. The Company's experiences in 1867 serve as a powerful reminder of the contingent and messy more-than-human geographies that shaped the RMSPC's operations. The mobility of the 1867 hurricane altered steamers' velocity and produced friction in the Company's network. Thus the 'constellation of mobility' of the service was revealed as fragile and reliant on a particular coalescence of non-human conditions (Cresswell 2010: 17).

Conclusion

At a meeting in October 1868 the RMSPC's chairman reflected upon the Company's service in the preceding years and stated:

> Whilst the Harbour of St Thomas was in ruins by the Hurricane, coals blown down, and people all at their wits end, then the Earthquake came and completed the destruction of the place. A panic seemed to seize the people, and perhaps not unnaturally, under the circumstances you will say, but it was most unfortunate, for everybody avoided St Thomas and the Yellow Fever, so that I very almost say that we had nature herself conspiring against us. (NMM RMS 4/4, 28 October 1868: 19–20)

This suggestion that 'nature' was conspiring against the Company underlines the important ways in which the RMSPC's fortunes were shaped by more-than-human as well as human factors.

The RMSPC's maritime operations were 'precariously vulnerable to non-human interventions' by weather systems, animals or organisms (Woods 2007: 498). Despite the Company's desire to provide a service characterized by predictability, order, and regularity, the steamship service was a contingent network of material, human and more-than-human elements that had to coalesce in particular ways to produce 'regular' journeys that might progress in accordance with the Company's scheme of routes. Through a view on the more-than-human world, and particularly moments of exception for the steamship service, I have suggested that the RMSPC's operations in ordinary and extraordinary moments were characterized by a 'multiplicity of space-times generated in/by the movements and rhythms of heterogeneous association' (Whatmore 2002: 6). In the maritime world, more-than-human considerations were constant. These had to be negotiated through alterations, adaptations and responses in everyday and exceptional moments, in order to achieve an elusive but highly desirable regularity of service.

Sources

National Maritime Museum, Greenwich, London

Archive of the Royal Mail Steam Packet Company.

(NMM RMS)

RMS 4/4 Verbatim reports of general meetings, 1862–1867.
RMS 6/3 In-letters, 1844–1845.
RMS 6/17 In-letters, 1867–1868.
RMS 7/1 Out-letters, 1839–1844.
RMS 7/2 Out-letters, 1844–1852.
RMS 36/4 Table of routes volume III.
RMS 38/1 RMSPC regulations, 1850.
RMS 54/2 Reports and Accounts, 1860–1869.

University College London Special Collections Royal Mail Steam Packet Company (UCL RMSP).

RMSP 13 Reports and accounts, 1842–1872.
RMSP 15 Cost and outfit of steamships, 1839–1912.

References

Anim-Addo, A. 2011. 'A wretched and slave-like mode of labour': Slavery, emancipation and the Royal Mail Steam Packet Company's coaling stations. *Historical Geography*, 39, 65–84.
Barad, K. 1998, Getting real: Technoscientific practices and the materialization of reality. *Differences: A Journal of Feminist Cultural Studies*, 10(2), 88–128.

Booker, J. 2007. *Maritime Quarantine: The British Experience, c. 1650–1900.* Farnham: Ashgate.

Braun, B. 2005. Environmental issues: Writing a more-than-human urban geography. *Progress in Human Geography*, 29(5), 635–50.

Bushell, T.A. 1939. *"Royal Mail": A Centenary History of the Royal Mail Line 1839–1939.* London: Trade and Travel Publications.

Castree, N. and Nash, C., 2006. Posthuman geographies. *Social and Cultural Geography*, 7(4), 501–4.

Coyle, F. 2006. Posthuman geographies? Biotechnology, nature and the demise of the autonomous human subject. *Social and Cultural Geography*, 7(4), 505–23.

Cresswell, T. 2010. Towards a politics of mobility. *Environment and Planning D: Society and Space*, 28(1), 17–31.

Curtin, P.D. 1998. *Disease and Empire: The Health of European Troops in the Conquest of Africa.* Cambridge: Cambridge University Press.

De Paolo, C. 2006. *Epidemic Disease and Human Understanding: A Historical Analysis of Scientific and Other Writings.* Jefferson, North Carolina: McFarland & Company.

Edensor, T. (ed.) 2010. *Geographies of Rhythm: Nature, Place, Mobilities and Bodies.* Farnham: Ashgate.

Gibbs, L.M. 2009. Water places: Cultural, social and more-than-human geographies of nature. *Scottish Geographical Journal*, 125(3–4), 361–9.

Hannam, K., Sheller, M., and Urry, J. 2006. Editorial: Mobilities, immobilities and moorings. *Mobilities*, 1(1), 1–22.

Harcourt, F. 2006. *Flagships of Imperialism: The P&O Company and the Politics of Empire From its Origins to 1867.* Manchester: Manchester University Press.

Hays, J.N. 2005. *Epidemics and Pandemics: Their Impacts on Human History.* Santa Barbara, California: ABC Clio.

Kingsley, C. 1874. *At Last: A Christmas in the West Indies.* London: Macmillan and Co.

Lambert, D. 2008. The "Glasgow king of Billingsgate": James MacQueen and an Atlantic proslavery network. *Slavery and Abolition*, 29(4), 389–413.

Lambert, D. and Lester, A. (eds), 2006. *Colonial Lives Across the British Empire: Imperial Careering in the Long Nineteenth Century.* Cambridge, Cambridge University Press.

Lambert, D., Martins, L. and Ogborn, M. 2006. Currents, visions and voyages: Historical geographies of the sea. *Journal of Historical Geography*, 32(3), 479–93.

Lorimer, J. 2010. Moving image methodologies for more-than-human geographies. *Cultural Geographies*, 17(2), 237–58.

MacQueen, J. 1838. *A General Plan for a Mail Communication by Steam between Great Britain and the Eastern and Western Parts of the World; also, to Canton and Sydney, Westward by the Pacific: to Which Are Added, Geographical Notices of the Isthmus of Panama, Nicaragua, &c. with Charts.* London: B. Fellowes.

Panelli, R. 2010. More-than-human social geographies: Posthuman and other possibilities. *Progress in Human Geography*, 34(1), 79–87.

Peters, K. 2012. Manipulating material hydro-worlds: Rethinking human and more-than-human relationality through off-shore radio piracy. *Environment and Planning A*, 44, 1241–54.

Richardson, B.C. 1997. *Economy and Environment in the Caribbean: Barbados and the Windwards in the Late 1800s*. Barbados: UWI Press.

Smith, C., Higginson, I. and Wolstenholme, P. 2003. 'Imitations of God's own works': Making trustworthy the ocean steamship. *History of Science*, 41(4), 379–426.

Smith, S.D. 2012. Storm hazard and slavery: The impact of the 1831 Great Hurricane on St Vincent', *Environment and History*, 18(1), 97–123.

Steel, F. 2011. *Oceania Under Steam: Sea Transport and the Cultures of Colonialism, c. 1870–1914*. Manchester: Manchester University Press.

Whatmore, S. 1999. Hybrid geographies: Rethinking the 'human' in human Geography. In *Human Geography Today*, edited by D. Massey, J. Allen and P. Sarre. Cambridge: Polity Press, 22–39.

Whatmore, S. 2002. *Hybrid Geographies: Natures Cultures Spaces*. London: Sage.

Woods, M. 2007. Engaging the global countryside: Globalization, hybridity and the reconstitution of rural place. *Progress in Human Geography*, 31(4), 485–507.

Chapter 12

Taking More-Than-Human Geographies to Sea: Ocean Natures and Offshore Radio Piracy

Kimberley Peters

Introduction

As the introduction to this book outlines, water worlds permeate our everyday existence in profound but often invisible ways. Over the course of 12 chapters, this volume has brought to the surface the importance of social, cultural and political geographers taking the seas and oceans seriously. These are not merely physical spaces of scientific interest, rather, *people* engage with the seas in all kinds of ways. They (we) cross the seas, map the seas, imagine the seas, use the seas and exploit the seas. As Steinberg confirms,

> [t]he ocean is not simply an environment wherein distinct marine phenomena may be observed by marine specialists, (i.e. marine geographers). Rather it is a space that, like land, shapes and is shaped by a host of physical *and* social processes. (1999: 367 emphasis added)

However, whilst over the past decade a plethora of work has begun to explore the human geographies of ocean space (see Lambert et al. 2006 and Peters 2010 for reviews) scholars have not comprehensively explored the sea as a *more-than-human* space; taking its very nature into account when examining social processes which occur, on and under its surface. In other words, there has not been sustained attention to both the 'physical *and* social' dimensions of ocean space and how the two are intimately woven together (Steinberg 1999: 367 emphasis added).

This is in spite of broader shifts within the discipline to consider nature-culture relations (see Castree and Braun 2001, Whatmore 1999), and unpack a singular view of nature as distinct from the social realm (Lorimer 2011: 197). Approaches which consider the more-than-human dimension of our world are currently at the forefront of academic endeavours as scholars explore the entanglements of the '(geo) earth and the bio (life)' (Whatmore 2006: 601) through studies of human engagements with coastal landscapes (Wylie 2005), weather worlds (Ingold 2008), gardens (Hitchings 2003); and moreover, a range of lives from the human to more-than-human plant life, animals (Bear and Eden 2010; Lorimer 2010, Philo and Wilbert 2000) and the lives of material things (Bennett 2004).

This absence of the sea as a more-than-human nature, co-fabricated with human life in such debates reflects the marginalization of water worlds in the discipline (see Introduction, this volume). Given the integral role water worlds in our everyday lives, it is important to take studies of the more-than-human to sea, 'looking outwards to watery horizons, considering how the 'hydro (sea)' and 'bio (life)" come together (Peters 2012: 1244). This means considering the water in and of itself as a more-than-human materiality; constituted of matter – particles through which energy travels and which sways and moves with a dynamism unlike the land – and that subsequently results in particular co-compositions when combined with human life (see also Peters 2012). Such an exploration of the sea as a more-than-human materiality is essential if the ocean is to be fully understood. As Lambert, Martins and Ogborn write, 'clearly climatic, geophysical and ecological processes belong in work on the sea' (2006: 482). Human geographers cannot then, ignore the ocean's nature.

In the two previous chapters (see Bear, and Anim-Addo), a 'non' and 'more-than-human' approach has been taken to the study of the sea, thinking through the agential impacts of a living world on human life. However, in both of these chapters, the sea as something more-than-human has been peripheral to the study of other more-than-human aspects present in the watery examples described (such as scallops and mosquitos). In this final chapter, I return to the sea more explicitly, as something which is beyond human construction – a physical, more-than-human space, with qualities and processes driven by the dynamism of the planet (and dynamism beyond the planet) (Clark 2011) – one which is entangled with socio-cultural and political *human* engagements leading to specific *socio-spatial* outcomes. I do this through drawing on a particular empirical example: pirate radio. I explore the difficulties faced by radio pirates in maintaining their operations at sea because of the mobile force of the sea and the impacts or *affects* of such physical nature on human life. In doing so, I contend that to fill the watery void in geographical research, we need to look to the water itself as a materiality which blends, merges and mixes with human life in significant ways. I begin this effort with an illustrative case study.

Offshore Radio Piracy

Narratives of the sea are appealing because they evoke a certain romantic sensibility (see Mack 2011: 23–35). This can be witnessed in the manifold literary and pictorial sources which focus attention on the maritime realm (from *Moby-Dick* to the paintings of J.M.W. Turner). The 'social construction of the ocean' in such sources (to follow Steinberg 2001) – as a space of the sublime and a space of wildness to be mastered by man [sic] – feeds our imagination. The story of pirate radio beginnings; of a 702 tonne ex-passenger ferry which was anchored three miles from the shore of Essex, outside UK territorial waters in the North Sea, converted to the radio station and playing everything from Cat Stevens to The Rolling Stones, is one which often sparks the imagination (Humphries 2003: 16). As disc-jockey Bill Legend stated,

> [I] discovered that by listening to the radio that there was some fantastic imagery available ... I discovered that this little box [a transistor radio] created some fantastic pictures and when I was thirteen, accidently I just hit [upon] this radio station which was just mind blowing, just phenomenal, they were operating off the coast of Holland – so I had all this imagery about these guys on [Radio] Caroline .... And it was just the most fantastic picture in my mind. (Interview June 2008)

However, whilst '[m]uch writing about the sea ... employs the sea as a metaphor rather than a lived reality' (Mack 2011: 25) the books, CDs and films charting the pirate radio story were not fictional accounts, but based on the actual lived realities of being at sea. Bill Legend's memories of listening to pirate radio (before he later became an offshore DJ himself) were not simply imaginings; they were built upon real human engagements with the water world.

The pirate radio phenomenon began on Easter Sunday of 1964, with a station named 'Radio Caroline' broadcasting from the aforementioned ship, the MV *Frederica*.[1] Offshore pirate radio was a direct response to the regulatory broadcasting climate in Britain at the time. Unlike America, where broadcasting had developed commercially with relatively few constraints (see Street 2002: 24), radio programming in Britain was subject to much tighter restrictions. Broadcasting was perceived to be a powerful medium which could be open to abuse. Its use, therefore, should be strictly regulated. Consequently, in 1922, six major radio manufacturers came together to form the British Broadcasting Company, which in 1927 would become the British Broadcasting *Corporation* (BBC), an independent, autonomous, but single unified broadcasting agency, endorsed by, but free from the controls of government (see Cain 1992). The BBC was the sole permitted broadcasting agency – leaving no legal scope for competition.

The BBC provided a mixed programming approach that was designed to appeal to the majority and minority of listeners, providing a variety of programmes including 'plays, concerts, charity appeals, debates, variety shows, weather forecasts, SOS messages, women's talks, dance band broadcasts, symphony concerts and religious talks' (Cain 1992: 13). However, in spite of a supposedly

---

1   The *Frederica* was the first ship to house Radio Caroline, but it was not the only ship used between 1964–1968. During this period Radio Caroline had two ships, one operating in the Irish Sea, broadcasting to the Isle of Man, Northern Ireland, Southern Scotland, North Wales and the North West (the *Frederica*), and the other airing transmissions from the North Sea (the *Mi Amigo*), broadcasting across South East England.

all-inclusive approach to broadcasting, the director-general, John Reith, had firm thoughts on the direction and purpose of programming, believing it to be 'the unquestioned duty of the educated and cultured elite to use broadcasting for further enlightenment of the public as a whole' (McDonnell 1991: 11). Programming was organized around themes of 'information', 'education' and 'entertainment' (Cain 1992: 12) with the audience and their tastes placed on a 'cultural ladder', with 'popular music' occupying the 'bottom rung' (Lewis and Booth 1989: 79).

With broader social, cultural, political and economic changes, by the 1960s, the BBC of 1927, with its 'public service' aims, was deemed to be out of touch with a more affluent, baby-boom generation, and an emerging 'youth' culture who were experiencing freedoms quite unlike those of any generation before (see Donnelly 2005, Marwick 1998). As McDonnell writes, 'a sizeable part of the audience was beginning to think that the BBC conception of service was too close to moral and cultural snobbery and too far removed from popular taste' (1991: 22). The BBC's highly principled programming meant they were both resistant to, and lacked the expertise in, playing the exciting new music which was shaping Britain's audio landscape (Barnard 1989: 17). As Lewis and Booth write,

> ... during the 1960s rock n'roll music became much more central to the existence of the young. ... Rock music became an all-pervasive accompaniment to the lives of young people whatever their class. It was woven into their work and their ideology as well as their leisure: it became the soundtracks for their lives. (1989: 81–2)

With limited airplay of rock 'n'roll music airing on the schedule of BBC outputs, entrepreneurs such as Ronan O'Rahilly (founder of Radio Caroline) recognised that by locating a station beyond the legal boundary of a nation, in international waters, and on board a ship,[2] British law would not apply and the British broadcasting monopoly could be evaded. Moreover, no international law would be broken either. At this time the 1958 United Nations Convention on the High Seas (otherwise known as the Geneva Convention) was in place. This did not 'explicitly include or exclude the freedom to broadcast at sea' (Robertson 1982: 79) and therefore broadcasting from the high seas was 'characterized as a freedom' of that space (Robertson 1982: 82). Using the medium-wave band frequency (MW), which allowed signal reception over long distances, listeners on land could tune into the latest music emanating from these extra-territorial sea-based stations. Such was the success of the first station, 'Caroline', that a further 14 sea-based stations emerged between 1964–1967 (Skues 2009: 89). In 1967 the British government introduced an 'anti-pirate' law, the Marine &etc. Broadcasting Offences Act (MBO Act), to counter broadcasting that was occurring

---

2   Or platform – some stations were based on old naval forts in the North Sea, for example Radio City.

beyond their control. This law worked to eradicate most radio pirates by cutting off the vital ship to shore links which provided stations and their staff with everything from water, food and fuel, to post, Mars Bars and vinyl records (see Lodge 2003). It also prevented British companies advertising on station broadcasts, effectively curtailing the revenue which made pirate enterprises possible. In spite of this law, radio piracy continued into the 1970s and broadcasting ships were only 'sunk' in 1991 with a new section of law, Schedule 16 of the Broadcasting Act, introduced on the 1 January that year.[3] The last station to broadcast, was the station which was the also the first[4] to broadcast, Radio Caroline. In this chapter I will be focusing solely on this particular station and its ship the MV *Amigo*.[5]

As I have noted elsewhere, the socio-cultural and political framing in which the pirate radio story sits, is not outside of a more-than-human context (Peters 2011, 2012). The sea was crucial in this story, and as I will show below, in particular to efforts by the pirates to maintain their activities. Indeed, what made the pirate radio narrative so appealing to many listeners was the romance of it as a maritime, sea-based phenomenon. As pirate radio listener Dean Masters told me, listening to the station was to listen to

> all of those who endured storms, tempests, shipwrecks, and dodged the authorities, just to have the freedom to play records to whoever chose to tune in and listen to them. (questionnaire response June 2008)

Here was not only a story of long enduring battle between evasive pirates and (largely) hostile governments – it was also a tale of a long enduring battle against currents, waves, tidal movements, storms – of attempts to master the more-than-human nature of the water, and instances of total surrender to the elements. This was a story of defiance again huge odds – human *and* physical – governmental *and* meteorological. The physical world of dynamic processes was at the very centre of this story. I next examine the ways in which water worlds, in and of themselves, were entangled with the socio-cultural *human* world in the case of Radio Caroline's offshore endeavours.

---

3   For a discussion of this, see Peters 2011. The MBO Act was limited because nation-states had no power to prevent the ships from operating at sea (a protected freedom under the 1958 Geneva Convention). They could only work to stop pirate radio stations by controlling parts of the enterprise that fell within their territorial borders. This gave pirate stations the license to continue.

4   Radio Caroline was the first station to broadcast specifically to a UK audience. There were, however, offshore pirate radio stations (such as Radio Veronica) which preceded Caroline, based in the Baltic Sea in the late 1950s (see Humphries 2003).

5   Radio Caroline's primary vessel 1964–1980. From 1983–present Radio Caroline's home is the MV *Ross Revenge* (when I say 'present', I refer to the fact that although the offshore pirate radio enterprise ceased in 1991, Radio Caroline maintains this final ship, making occasional anniversary broadcasts from Tilbury dockyard, where the ship resides).

The Co-Fabrication of Fluid Nature and Human Life

The primary way in which radio pirates sought to circumnavigate the anti-pirate law, the 1967 MBO Act, was simply to stay at sea. As noted earlier, radio pirates were located in international waters, beyond British jurisdiction. Therefore their activities could not be held to account nationally, nor were they accountable internationally, because there was no stipulation against offshore broadcasting outlined in the Geneva High Seas Convention of 1958.[6] Staying at sea meant that ships (and their crews) were protected from any legislation which may impact them, should they return to British shores. Evasion did not depend on mobility, but rather on stillness and immobility (Cresswell 2012): on staying in position, whatever the odds.

Maintaining the pirate enterprise then, meant being resolutely fixed at sea. Indeed, unlike typical ships which travel from place A to B, pirate radio ships did not go anywhere. Ships were anchored to the seabed in one particular location for months, sometimes years at a time. As Jason, an engineer on board the Radio Caroline ship MV *Mi Amigo* (1978–1980) told me during one interview,

> Yeah, most people who um, go to sea, like the merchant navy, royal navy, or just generally, go somewhere – they leave London they go to the Far East, America, places like that, with Caroline, you go somewhere and stay there and you don't go anywhere else. So it's like being dumped on an err.. desert island really (interview June 2008).

Whilst legally it was essential that the ship was as static as possible, there were other reasons why it was beneficial to keep a steady location. Firstly it permitted the best possible frequency reception for listeners on land. Those operating radio stations could select desirable distances which would permit the optimum reception to intended audiences in Britain (Skues 2009). Secondly it ensured that tender boats, which supplied the radio ships with necessary amenities, knew where to find the *Mi Amigo* (Burt Johnson Interview May 2009). If the radio ships were to move, it would be more difficult for 'tenders' (as they were called) to locate ships and drop off supplies. Furthermore, and most significantly, those running pirate radio stations were acutely aware that the operation would be at the whim at nature. Bad weather might force transmissions to close down, and could threaten the ship and the lives of the crew. Staying in one location allowed pirate radio bosses to select sites where the more-than-human materiality of the sea, and the

---

6   It should be noted, however, that pirates did break the regulations of the International Telecommunications Union because they were broadcasting on unallocated (and therefore unauthorized) frequencies. Yet, because it is the responsibility of nations to adjudicate matters impacting their territories, and because the radio pirates were beyond their territory, this made regulation for British authorities inherently problematic.

forces running through and on it from jet streams, winds and gravitational pull, would be most manageable in view of maintaining the enterprise.

Indeed, it would often be the case that life on board was impacted by the movement of the surface of the sea (and the subsequent undulation of the ship on that motionful surface). Most insignificantly, records would skip (Blackburn 2007: 75), more significantly, disc-jockeys (DJs for short) and other members of crew would experience sea-sickness (Walker 2007: 69, Interviews 2008–2011). It was in the best interests of the management to select appropriate anchorages where these impacts would be minimized. For the most part,[7] from 1974–1987, Radio Caroline was anchored at Knock Deep, an area located 51°35'00"N 01°17'20"E in the English Channel. Knock Deep is a particular area littered with sandbanks. Anchoring here meant that the Caroline ships could be sheltered from the movement of the sea, as the shallower areas of sandbanks lessened the swell, protecting the vessel from the full force of the North Sea. As DJ Johnnie Walker wrote of his arrival to the MV *Mi Amigo*,

> There was a gentle rolling swell that lessened as we got near the home of Radio Caroline. Ronan [O'Rahilly, Caroline's founder], on his rare visits to the ship, used to tell visitors and journalists it was the calming effect of The Lady, as he referred to Caroline. His was a more romantic and mystical explanation than the fact that the ship was moored over the Knock Deep sandbank and the seas were always calmer there. (2007: 88)

However, in spite of such planning to fix the ship securely in place, in an area of sea known for calmer conditions, it remained the case that the ship moved; be it with turning tides (interview, Ray Clark 2009) or with the gentle roll of waves (Walker 2007: 88). As Bear and Eden note (regarding the role of fish certifications, 2008) trying to fix anything at sea is impossible – the sea moves (and in their case the fish move too, 2008: 495). Being liquid,[8] the sea has a composition of particles which means it moves in relation to terrestrial forces (such as the wind) and extra-terrestrial gravitational forces (the pull of the moon). It is a dynamic, motionful, three-dimensional space because of its physical nature. As such, the effort to fix the Caroline ships in place, however strategically, was always destined to fail. The maintenance of the enterprise at sea was always going to be difficult because it was *at sea*, subject to forces beyond human control. This failure to remain fixed was drastically felt on a number of occasions in Radio Caroline's sea-based history and it resulted in various affects for those working onboard. In the vignette which follows, I focus on one notable occasion.

---

7 Aside from driftings, and the period 1980–1983 when there were no offshore pirates broadcasting to the UK.

8 In this instance – see Vannini and Taggart, in this collection, for the sea as a solid.

November 1975

In spite of the protected position the Knock Deep anchorage provided, the autumn of 1975 was an eventful one for Radio Caroline. Keeping both the ship *Mi Amigo* and the crew on board as fixed and stable as possible had been largely problematic. It had been a particularly rough period at sea. In the September, crew on board the *Mi Amigo* had witnessed a nearby vessel in distress, firing emergency flares (Humphries 2003: 135). Little over two months later it would be the *Mi Amigo* in distress as 'sea conditions worsened' (Humphries 2003: 135). This culminated with a storm, which hit the position of the vessel on Saturday 8 November. Home Office (HO) records reported 'the wind was a NE Force 8, gusting to Force 9' (HO 255/1219). These 'severe gales' continued through to the 9 November, by which time, the winds had caused 'very rough', 'high' waves (Beaufort Wind Force Scale 2013), forcing the ship *Mi Amigo* from the security of its anchorage.

The *Mi Amigo* was not a large ship. At only 135ft long, and 470 tonnes, Peter Moore, Radio Caroline's manager, described the vessel as 'a fragile little thing' (Interview July 2008). In bad weather the waves would literally ride over the ship because it sat so low in the water (Peter Moore Interview July 2008). It was a 'tiny' ship which DJ Tony Blackburn stated 'swung like a pendulum' when the waves got rough (2007: 69, 75). On the 8 November, the weather conditions impacted the sea conditions, causing the radio ship to '[snatch] violently at its anchor chain' (Humphries 2003: 135). This was because the wind direction was driving waves towards the vessel from the only direction 'which the ship was not protected' by the shallower sea and sandbanks (Humphries 2003: 135). The Home Office, who were surveilling the vessel under the MBO Act, were keeping a close eye on the situation and reported that, as conditions worsened,

> the *Mi Amigo's* anchor chain broke. … The ship drifted southwards over the Long Sands … the ship then appeared to travel under power in a southward direction until it was 8 miles north of Margate. … The engines then failed and at about 23 00hrs, it drifted westwards with the wind and eventually stopped at position 51°3'42 N 01°15'18'E. (HO 255/1219)

During this period, the ship and its crew were hostage to the coalescing forces of nature; a world in a constant state of elemental flux and becoming. Here the solidity of the ship, the liquidity of water, and the force of the wind came together in a fluid assemblage impacting the lives of those onboard. The efforts of the crew to immobilize themselves with an anchor chain were futile amidst a severely motionful, undulating sea, whose movement (driven by the gale force winds) was no match for such strategies. Accordingly the ship began to move too, becoming disconnected from the seabed, drifting uncertainly. The affects for those on board ranged from panic, desperation to confusion (Humphries 2003: 135). Humphries notes how there was a 'fear' amongst the crew on board (2003: 135–6). Once adrift, the crew were also disorientated without navigational equipment, and had no idea of their location and whether they were inside or outside British territorial waters (Noakes 1984: 236). As the passage above describes, the crew reacted, trying to take control of the situation, powering the vessel in response to the movement induced by the severe storm conditions and the dynamic nature of sea; yet this was to no avail.

The mobility of the ship, resulting from the more-than-human nature of the sea itself (a physical body of particles held together less tightly than those of other natures, thus subject to flex and motion when energy is transmitted through it) had direct impacts on those living and working in such a distinct environment. The combination of the sea and the broader weather world (see Ingold 2008) had effectively prevented radio pirates from maintaining their broadcasting activities. Over the coming week, the Radio Caroline organization attempted to rescue and secure the drifting ship (which had strayed into British waters and was no longer broadcasting). Members of the organization rallied together, first sending out a 'fast, twin prop tender' to take the *Mi Amigo* and crew 'in tow in a northwest direction' to secure a new anchorage (HO 255/1219). This was unsuccessful. The next attempt was no better, as when '[t]he supply vessel made several attempts to tow the *Mi Amigo* ... damage was caused to the radio ship, and part of the ship's side rail was ripped off' (Humphries 2003: 137). In spite of the failure of these efforts of the crew and wider Caroline community to re-stabilize the *Mi Amigo*, it demonstrated the impacts the weather had on the sea and which the sea had on the social organization of the enterprise. People had to pull together in ways they may not have ordinarily because of the unpredictable and sometimes dangerous nature of living upon the ocean. The motion of the sea was part of the assemblage which influenced social life. As Bill and Jason, crew members on the *Mi Amigo* during the 1970s mused,

> J: the anchor chain [was] twisted and could break at any moment, lots of plates leak[ed] underneath [the ship] and there was always the danger it might drift.

> B: But ... it was a commune ... you feel an important part of the cog ... You have to get on with people cos ... you were there to do a job and [it could be] very, very scary at times. (Interview June 2008)

As Jason and Bill described, the 'scary' nature of living at sea, with threats of twisted chains which could mobilize the ship from its secure anchorage, meant that crews took on a commune-like social structure, working together to battle against the forces which undermined their efforts.

In 1975, the *Mi Amigo* remained adrift until 26 November. Only then did it return to the 'proper anchorage' of Knock Deep, to be immobilized and secured once more in international waters, able to broadcast to eager listeners on land (Humphries 2003: 138). For over two weeks the more-than-human, three-dimensional materiality of the sea had threatened the maintenance of the enterprise and even led to a governmental raid on 14 November, whilst the vessel was drifting in British territorial waters (HO 255/1219). Such meteorological conditions affected the crew and the wider community involved in the pirate radio enterprise, whose lives were relationally affected as the sea was subject to these intermingled conditions driven by the weather. Accordingly, as Steinberg notes, 'physical *and* social' processes come together in water worlds (1999: 367).

Moreover, these processes are distinct at sea, where water (and its physical composition) is concerned. Whatmore notes that the contribution of more-than-human approaches is to 're-animate the missing 'matter' of landscape, focusing [our] attention on bodily involvements in the world in which landscapes are co-fabricated between more-than-human bodies and a lively earth' (2006: 603). However, these 'involvements' with the world around us have focused largely on the 'matter' of the terrestrial or 'earthy' sphere, rather than the materialism of the watery world (as indicated and enforced by the use of language: '*land*scape' as opposed to '*sea*scape' for example). Yet the sea is not the land, its more-than-human nature, is fluid. Thus the when considering water worlds, alternative co-compositions between the physical and social arise.

Indeed, the maintenance of the pirate radio enterprise, a human, socio-cultural and political endeavour, relied, oddly, on the ship being immobile, staying securely anchored in one place. This meant countering the forces of a distinctly fluid nature. Yet as the vignette demonstrated, the physical motion – the forcefulness of the ocean – driven by the assembled forcefulness of the wind and gravitational pull – made this maintenance problematic. The ship was never still: even when anchored, it moved. In the worst cases, the ship would be cast adrift.[9] Life onboard the ship was intrinsically shaped by this fluid nature. As I have noted elsewhere (2012: 1252–3) '[h]umans cannot force power back on to the sea; shaping nature ... they can but harness its qualities, or manipulate [it] to best effect'. In other words, human life is always subject to the force of the sea, even as we try to lessen those forces. Unlike other more-than-human natures, when co-composed with human life, the sea sparks particular outcomes, which alter the relational balance between the social and the physical. Whereas humans might have greater agency over nature on land, this dynamic alters with the alternative material composition of the sea (Peters 2012), and accordingly humans are uniquely affected by it. As the vignette demonstrated, the sea evokes actions and reactions which shapes human interactions with the water world. The natural environment at sea, resulted in particular socio-cultural formations for the crew as they struggled to survive. The sea as a material, more-than-human nature, must therefore be at the forefront of our understandings as we grapple to understand human engagements with it.

9　This happened multiple times in Radio Caroline's history. Other notable dates are 1980, when the *Mi Amigo* again broke anchor, this time leading to the sinking of the ship, and again in 1991 when the ship drifted to the Goodwin Sands, an area renowned for ship wrecks, resulting in the rescue of the crew by helicopter (see Conway 2009, Humphries 2003 and Skues 2009 for detailed accounts).

More-than-Human Geographies *of* the Sea

In this chapter I have followed recent lines of enquiry, driven by a '*re*turn' towards a world of 'matter' – the very substance of the world around us – (Whatmore 2006: 603) to consider how more-than-human material natures are co-fabricated with human existence. Of late, geographers have become increasingly interested in this co-composition of human and more-than-human worlds (see for example social and cultural geographies of climate change, Brace and Geoghegan 2010 and responses to natural disasters, Clark 2011). This has followed a recognition that we need to consider a range of more-than-human or physical attributes which characterize our socio-cultural world, the groundedness of the earth, animals, plant life, and 'things' we humans make (from trainers, to motor cars, to buildings) in order to move away from the bias on understanding human agency as *the* dominant agency in the world in which we live (see Whatmore 2006: 603). As Bennett puts it, there are 'things' in our world which contain their own power, without humans assigning that power or significance to them (2004: 48). As she deftly puts it: 'there is an existence peculiar to a thing that is irreducible to the thing's imbrication with human subjectivity' (Bennett 2004: 348). In other words, the more-than-human approach demands we consider the power of material things; objects, natures, and so on, to avoid the '[o]veremphasis on human agency' in composing the world in which we live (Lambert et al. 2006: 482).

For Bennett (2004), Whatmore (2002) and others (Bear and Eden 2008, Clark 2011, Hitchings 2003, Jones 2011 and Philo and Wilbert 2000) this marks a shift in social and political thought which considers carefully the vibrancy of the planet – that there is, in Bennett's words, a world of 'powerful things' which have dynamism or *force* without human's assigning them such qualities (2004: 348). This, as Thrift argues, opens up 'a different kind of intelligence about the world' (2004: 60). Such a project,

> (fosters) greater recognition of the agential powers of natural.. things, greater awareness of the dense web of their connections with each other and with human bodies. (Bennett 2004: 349)

Such an awareness or 'recognition' of the powers of nature, demands that we, (humans),

> [t]read lightly upon the earth, both because things are alive and have value as such *and* because we should be cautious around things that have the power to do us harm. (2004: 365–6)

The natural more-than-human world, Bennett reminds us, must be acknowledged. It has an agency, or power, which is often beyond our control. It can impact and shape our lives (as much as, or more than, we can impose power upon it; see Hitchings 2003, Peters 2012, Whatmore 2006).

However, it is the power dynamics of more-than-human natures which is called into question when we consider the fluid materiality of the sea, compared to the earthy solidity of land. As I have argued elsewhere (Peters 2012) the specific material and elemental quality of the sea itself, its composition of particles, its relation to wider forces of power, means that it becomes a distinct space for human engagement; one we can move on, in and through; one which itself moves, and one which shifts states (solid, liquid to air). Human life does not play out in water worlds in the same way it does on landed worlds (Peters 2012: 1252). As such, looking to the sea, and the corporeal experiences and sensibilities which are made possible in these spaces, enables us to grasp new knowledge about the world.

I have considered the sea itself as a more-than-human nature; a material entity through which force operates and human affects are elicited. Through the case study vignette, I have filled the liquid void in geographies of the seas *with the water itself*, thinking seriously about its composition and the impacts of force on its structure, which relate to and are co-fabricated with the human worlds on board pirate radio vessels. Thinking about the events of November 1975 in a more-than-human framework allows 'greater recognition of the agential power of natural ... things' (Bennett 2004: 34). It has made possible an examination of how the physical and socio-cultural world interrelate, to unpack the particular socio-spatial outcomes which result when humans engage with the fluid nature of water.

Conclusions

In the foreword to this book, Philip Steinberg subverted the 70% statistic often used to justify sea-based studies. However, he interestingly noted that one reason this is such a 'compelling figure' (Steinberg, this collection) is because 70% of the human body is also water. The much cited 70% then, has significance in both the human and natural world, and shows a synergy between both human and water worlds. We are, in the words of Margulis and Sagan, 'walking, talking minerals'; walking, talking *liquids* (1995: 49, in Bennett 2004: 360). It makes logical sense then, that studies of the water are not marginal to those of human *being* and accordingly, one way to think about human geographies of the water worlds is to take the water itself seriously, and moreover, to think about those water worlds fluidly, as open, permeable and subject to change; co-constituted and co-fabricated through broader relational elemental and human assemblages.

In this chapter I have explored how the hydrological and human come together and the affects borne from these forceful combinations through focusing on the sea itself as a lively, relational, more-than-human materiality. I have demonstrated that extending more-than-human geographies from the shore, is one way in which human geographers may begin to take seriously the agency and productivity of the *dominant* natural feature in world and its importance in the making of human experience: the water. However, such a move has depended upon shifts in the discipline which have been directed towards the 'processes and excesses of 'livingness' in a more than human world' (Whatmore 2006: 604). In other words, to echo Steinberg in the Foreword of this book, scholars are now in a unique position to examine water worlds because we have the conceptual and theoretical tools, such as more-than-human approaches, to do so. Yet arguably, as this chapter has shown, we still need to take these tools to sea. More-than-human approaches have remained focused on the nexus between *land*scapes and bodies rather than *sea*scapes and bodies (Whatmore 2006). The language of the more-than-human then, has been 'landed', rather than fluid.

In the introduction to this volume, Jon Anderson and I argued for the need to develop a language which can make fluid worlds comprehensible and 'known'; a language which might also be applied more broadly; a fluid ontology for which the water world frames a wider world of flux, change, rhythm, movement and flow. However, whilst we may now have the tools available (such as more-than-human approaches) to enable us to start from the water and look outwards, we are still tasked with doing so. This task ultimately requires us to *unearth* our language. In this collection, we have set sail on such task, envisioning a *fluid* world in which water is an equally important part of a wider geographical, elemental and corporeal assemblage of land, air, human and more-than-human life, which is in a constant state of becoming.

Sources

National Archive Kew

Home Office records relating to the Radio Regulatory Department (HO) HO 255/1219.

References

Barnard, S. 1989. *On the Radio: Music Radio in Britain*. Milton Keynes: Open University Press.

Bear, C. and Eden, S. 2008. Making space for fish: The regional, networked and fluid spaces of fisheries certification. *Social and Cultural Geography*, 9(5), 487–504.

Bennett, J. 2004. The force of things: Steps towards an ecology of matter. *Political Theory*, 32(3), 347–72.

Blackburn, T. 2007. *Poptastic! My Life in Radio*. London: Cassell Illustrated.

Brace, C. and Geoghegan, H. 2010 Human geographies of climate change: landscape, temporarily, and lay knowledges. *Progress in Human Geography*, 35(3), 284–302.

Cain, J. 1992. *The BBC: 70 Years of Broadcasting*. London: British Broadcasting Corporation.

Castree, N. and Braun B. 2001. *Social Nature: Theory, Practice and Politics.* Oxford: Wiley.

Clark, N. 2011. *Inhuman Nature: Sociable Life on a Dynamic Planet.* London: Sage.

Conway, S. 2009. *Shiprocked: Life on the Waves with Radio Caroline.* Dublin: Liberties Press.

Cresswell, T. 2012. Mobilities II: Still. *Progress in Human Geography*, 36(5), 645–53.

Donnelly, M. 2005. *Sixties Britain.* Harlow: Pearson Education Limited.

Hitchings, R. 2003. People, plants and performance: On actor-network theory and the material pleasures of the private garden. *Social and Cultural Geography*, 4, 99–113.

Humphries, R.C. 2003. *Radio Caroline: The Pirate Years.* Yately: The Oakwood Press.

Ingold, T. 2008. Bindings against boundaries: Entanglements of life in an open world. *Environment and Planning A*, 40, 1796–810.

Jones, O. 2011. Lunar-solar rhythmpatterns: Towards the material cultures of tides. *Environment and Planning A*, 43, 2285–303.

Lambert, D., Martins, L. and Ogborn, M. 2006. Currents, visions and voyages: historical geographies of the sea. *Journal of Historical Geography*, 32(3), 479–93.

Lewis, P.M. and Booth, J. 1989. *The Invisible Medium: Public, Commercial and Community Radio.* China: Macmillan.

Lodge, T. 2003. *The Ship That Rocked the World: The Radio Caroline Story: From the Inside as told by Tom Lodge.* California: UMI Foundation.

Lorimer, J. 2010. Moving image methodologies for more-than-human geographies. *Cultural Geographies*, 17(2), 237–58.

Lorimer, J. 2011. Nature. In *The Wiley-Blackwell Companion to Human Geography*, edited by J. Agnew, and J.S. Duncan. Oxford: Blackwell, 197–208.

Mack, J. 2011. *The Sea: A Cultural History.* London: Reaktion.

Marwick, A. 1998. *The Sixties: Cultural Revolution in Britain, France, Italy and the United States 1958–1974.* Oxford: Oxford University Press.

McDonnell, J. 1991. *Public Service Broadcasting: A Reader.* London: Routledge.

Noakes, B. 1984. *Last of the Pirates: A Saga of Everyday Life on Board Radio Caroline.* Edinburgh: Paul Harris Publishing.

Peters, K. 2010. Future promises for contemporary social and cultural geographies of the sea. *Geography Compass*, 4(9), 1260–72.

Peters, K. 2011. Sinking the radio 'pirates': Exploring British strategies of governance in the North Sea, 1964–1991. *Area* 43(3), 281–7.

Peters, K. 2012. Manipulating material hydro-worlds: Rethinking human and more-than-human relationality through off-shore radio piracy. *Environment and Planning A*, 44, 1241–54.

Philo, C. and Wilbert, R. 2000. *Animal Spaces, Beastly Places: New Geographies of Human-Animal Relations.* London: Routledge.

Robertson, H.B. 1982. The suppression of pirate broadcasting: A test case of the international system of control of activities outside National Territory. *Law and Contemporary Problems*, 45(1), 71–101.

Skues, K. 2009. *Pop Went the Pirates II*. Norfolk: Lambs' Meadow Publications.

Steinberg, P.E. 1999. Navigating to multiple horizons: Towards a geography of ocean space. *Professional Geographer*, 51(3), 366–75.

Steinberg, P.E. 2001. *The Social Construction of the Ocean*. Cambridge: Cambridge University Press.

Street, S. 2002. *A Concise History of British Radio 1922–2002*. Devon: Kelly Publications.

Thrift, N. 2004. Intensities of feeling: Towards a politics of affect. *Geografiska Annaler B*, 86(1), 57–78.

Walker, J. 2007. *The Autobiography*. London: Penguin Books.

Wylie, J. 2005. A single day's walking: Narrating self and landscape on the South West Coast Path. *Transactions of the Institute of British Geographers*, 30(2), 234–47.

Whatmore, S. 1999. Hybrid geographies: Rethinking the 'human' in human geography. In *Human Geography Today*, edited by D. Massey, J. Allen and P. Sarre. Cambridge: Polity Press, 22–39.

Whatmore, S. 2006. Materialist returns: Practising cultural geography in and for a more-than-human world. *Cultural Geographies*, 13, 600–609.

Index

Lightning Source UK Ltd.
Milton Keynes UK
UKOW06n1942160916

283181UK00014B/105/P